WATER RESOURCE PLANNING, DEVELOPMENT AND MANAGEMENT

WATER RESOURCES IN THE UNITED STATES

SELECT ISSUES

WATER RESOURCE PLANNING, DEVELOPMENT AND MANAGEMENT

Additional books in this series can be found on Nova's website under the Series tab.

Additional e-books in this series can be found on Nova's website under the e-book tab.

WATER RESOURCE PLANNING, DEVELOPMENT AND MANAGEMENT

WATER RESOURCES IN THE UNITED STATES

SELECT ISSUES

DIANNA HARRINGTON
EDITOR

NOVINKA
New York

NOTICE TO THE READER

Library of Congress Cataloging-in-Publication Data

ISBN: 978-1-63482-681-5

Published by Nova Science Publishers, Inc. † New York

CONTENTS

Preface		**vii**
Chapter 1	Water Resource Issues in the 114th Congress *Betsy A. Cody, Charles V. Stern, Nicole T. Carter and Pervaze A. Sheikh*	**1**
Chapter 2	U.S.-Mexico Water Sharing: Background and Recent Developments *Nicole T. Carter, Clare Ribando Seelke and Daniel T. Shedd*	**19**
Chapter 3	Drought in the United States: Causes and Current Understanding *Peter Folger and Betsy A. Cody*	**45**
Chapter 4	Emergency Water Assistance during Drought: Federal Non-Agricultural Programs *Nicole T. Carter, Tadlock Cowan and Joanna Barrett*	**69**
Index		**87**

PREFACE

This book provides an overview of the federal role in water resources development, management, and protection, with a focus on projects of the two major federal water resources agencies— Reclamation and the U.S. Army Corps—and related legislation. It also discusses overarching policy issues, such as drought and flood management and response, project funding and authorization priorities, and aquatic ecosystem restoration.

Chapter 1 – The 114th Congress faces many water resource development, management, and protection issues. Congressional actions shape reinvestment in aging federal infrastructure (e.g., dams, locks, and levees) and federal and nonfederal investment in new infrastructure, such as water supply augmentation, hydropower projects, navigation improvements, and efforts to restore aquatic ecosystems. These issues often arise at the regional or local levels but frequently have a federal connection. Ongoing issues include competition over water, drought and flood responses and policies, competitiveness and efficiency of U.S. harbors and waterways, and innovative and alternative financing approaches. The 114th Congress also may continue oversight of operations of federal infrastructure during drought and low-flow conditions, past large-scale flooding issues (e.g., Hurricane Sandy, Hurricane Katrina, Missouri and Mississippi River floods), and balancing hydropower generation, recreational use, and protection of threatened and endangered species. In addition to oversight, each Congress also provides appropriations for major federal water resource agencies, such as the U.S. Army Corps of Engineers (Corps) and the Bureau of Reclamation (Reclamation).

This report discusses recent congressional activity and possible topics for the 114th Congress. It provides an overview of the federal role in water resources development, management, and protection, with a focus on projects of the two major federal water resources agencies—Reclamation and the U.S.

Army Corps—and related legislation. It also discusses overarching policy issues, such as drought and flood management and response, project funding and authorization priorities, and aquatic ecosystem restoration.

Chapter 2 –The United States and Mexico share the Colorado River and Rio Grande pursuant to binational agreements. Compliance with these agreements becomes more complicated and controversial as water demands near or exceed available supplies and when drought and high heat further reduce availability and increase demand. This report describes legal obligations, drought conditions, and water sharing issues for this southwest region.

Chapter 3 – This report discusses how drought is defined (e.g., why drought in one region of the country is different from drought in another region), and why drought occurs in the United States. How droughts are classified, and what is meant by moderate, severe, and extreme drought classifications, are also discussed. The report briefly describes periods of drought in the country's past that equaled or exceeded drought conditions experienced during the 20th century. This is followed by a discussion of the nature and extent of recent droughts that affected Texas and the U.S. midcontinent, and the current drought in California. Lastly, the report discusses future prospects for a climate in the western United States that might be drier than the average 20th-century climate and the possible influence of human-induced climate change.

Chapter 4 – Drought conditions often fuel congressional interest in federal assistance. While drought planning and preparedness are largely individual, business, local, and state responsibilities, some federal assistance is available to mitigate drought impacts. While much of the federal assistance is targeted at mitigating impacts on the agricultural economy, other federal programs are authorized to provide non-agricultural water assistance. Interest in these non-agricultural programs often increases as communities, households, and businesses experience shrinking and less reliable water supplies. Authorized federal assistance is spread across a variety of agencies, and each has limitations on what activities and entities are eligible and the funding that is available.

In: Water Resources in the United States
Editor: Dianna Harrington
ISBN: 978-1-63482-681-5

Chapter 1

WATER RESOURCE ISSUES IN THE 114TH CONGRESS*

Betsy A. Cody, Charles V. Stern, Nicole T. Carter and Pervaze A. Sheikh

SUMMARY

The 114th Congress faces many water resource development, management, and protection issues. Congressional actions shape reinvestment in aging federal infrastructure (e.g., dams, locks, and levees) and federal and nonfederal investment in new infrastructure, such as water supply augmentation, hydropower projects, navigation improvements, and efforts to restore aquatic ecosystems. These issues often arise at the regional or local levels but frequently have a federal connection. Ongoing issues include competition over water, drought and flood responses and policies, competitiveness and efficiency of U.S. harbors and waterways, and innovative and alternative financing approaches. The 114th Congress also may continue oversight of operations of federal infrastructure during drought and low-flow conditions, past large-scale flooding issues (e.g., Hurricane Sandy, Hurricane Katrina, Missouri and Mississippi River floods), and balancing hydropower generation, recreational use, and protection of threatened and

* This is an edited, reformatted and augmented version of a Congressional Research Service publication R43910, prepared for Members and Committees of Congress, dated February 11, 2015.

endangered species. In addition to oversight, each Congress also provides appropriations for major federal water resource agencies, such as the U.S. Army Corps of Engineers (Corps) and the Bureau of Reclamation (Reclamation).

The issues before the 114th Congress are shaped in part by what earlier Congresses chose to enact and consider. Measures considered but not enacted by the 113th Congress include California drought legislation, various drought policy and water efficiency and conservation measures, regional restoration legislation (e.g., Klamath Basin, Great Lakes, Chesapeake Bay), actions to expedite water storage projects and permits, settlement of Indian water rights claims, and a lifting of restrictions on firearms at Army Corps projects.

Because of recent water conditions, disasters, or legal or agency developments, certain river basin issues are particularly likely to receive congressional attention during the 114th Congress. The Columbia River, Missouri River, and Sacramento and San Joaquin River (Central Valley Project) basins fall into this category. Other potential topics of congressional interest include emergency drought or flood legislation, private and public hydropower, water research and science investment and coordination, aging infrastructure, and environmental policy.

The 113th Congress enacted an omnibus Corps authorization bill, the Water Resources Reform and Development Act of 2014 (WRRDA 2014, P.L. 113-121). In addition to authorizing new programs (e.g., Water Infrastructure Finance and Innovation Act) and Corps construction projects, the legislation also established new processes that may shape how subsequent Corps project authorizations are identified. A Corps authorization bill often is considered by each Congress; enactment, however, has been less regular, with the most recent bills enacted in 2014, 2007, and 2000. The 113th Congress also enacted legislation to facilitate small conduit hydropower development (P.L. 113-23 and P.L. 113-24).

This report discusses recent congressional activity and possible topics for the 114th Congress. It provides an overview of the federal role in water resources development, management, and protection, with a focus on projects of the two major federal water resources agencies—Reclamation and the U.S. Army Corps—and related legislation. It also discusses overarching policy issues, such as drought and flood management and response, project funding and authorization priorities, and aquatic ecosystem restoration.

INTRODUCTION

The 114th Congress is likely to face numerous water resource issues as it conducts oversight and deliberates on authorizations and appropriations related to federal water resource development, management, and protection. Such issues include how to make investment decisions in the face of fiscal constraints; how to maintain and reinvest in an aging portfolio of federal infrastructure (e.g., dams, locks, and levees); how to effectively respond to and prepare for flood and drought emergencies; and how to distribute investment between activities to meet new demands for water supplies, navigation, flood management, and aquatic ecosystem restoration and protection. These issues often arise at the regional level but have a federal connection. For example, Congress may be faced with responding to various water-related crises, such as extreme drought or flooding issues (e.g., California drought in 2014 and coastal flooding issues associated with Hurricane Sandy or other storms). More broadly, Congress may be faced with addressing navigation challenges due to drought-induced low river flows or floods and water supply needs of farm and urban communities while also protecting threatened and endangered species. The crux of many of these challenges is how to balance competing demands for water and river management, including how to cope with the growing budget limitations and the effect of federal project operations on the environment.

This report first discusses recent congressional activity and possible topics for the 114th Congress. Next, it provides an overview of the federal role in water resources development, management, and protection, including a discussion of the two major federal water resources agencies—the U.S. Army Corps of Engineers (Corps) and the Bureau of Reclamation (Reclamation)—and related legislation. The report then provides an overview of overarching policy issues, including flood and drought preparedness and response, project funding and authorization priorities, and aquatic ecosystem restoration.

RECENT CONGRESSIONAL ACTIVITY AND LOOKING FORWARD

The water resource issues of the 114th Congress are shaped in part by the actions of past Congresses. Legislative activity often is specific to the federal water resource management agencies, such as the U.S. Army Corps of

Engineers in the Department of Defense and the Department of the Interior's Bureau of Reclamation, or it is specific to water use by particular sectors, such as energy, agriculture, navigation, recreation, and municipal and industrial use. Occasionally, Congress takes up broader water resource policy issues, such as coordination of federal water resource activities and programs. Legislation enacted for both the Corps and Reclamation during the 113th Congress was less than in prior Congresses, in large part due to congressional earmark policies, which may limit authorization of and appropriations for site-specific projects;[1] however, some broad legislation was enacted for both agencies.

Legislation Enacted in the 113th Congress

The 113th Congress enacted an omnibus Corps authorization bill, the Water Resources Reform and Development Act of 2014 (WRRDA 2014, P.L. 113-121).[2] Congress also included in separate legislation (P.L. 113-295) a provision increasing the fuel tax on commercial barges on federal inland waterways.[3] In addition, the 113th Congress provided regular and supplemental appropriations for the Corps to conduct its work and performed oversight on its flood and drought management and navigation actions, among other activities. The 113th Congress also provided appropriations for Reclamation to conduct its ongoing activities.[4]

Energy and environmental policy affect water resources management and development. Two bills intending to facilitate the development of nonfederal hydropower were enacted in the 113th Congress: a small conduit hydropower development bill to facilitate nonfederal hydropower development at Reclamation facilities (P.L. 113-24) and a separate bill making alterations to the Federal Energy Regulatory Commission licensing process for certain projects (P.L. 113-23). Several provisions related to water management in California were included in enactment of the FY2014 Consolidated Appropriations Act (P.L. 113-76, H.R. 3547) in January 2014. Also included in P.L. 113-76 was a one-year extension of the CALFED authorization (§207), and reauthorization of the Reclamation States Emergency Drought Relief Act of 1991 (extended through 2017; §206). The 113th Congress also reauthorized the National Integrated Drought Information System (NIDIS, P.L. 113-86), which, among other research, produces various Drought Monitor products.

The 113th Congress also enacted a farm bill (P.L. 113-79).[5] In addition to providing support for farmers and crop production, farm bills provide support for agricultural water conservation and efficiency measures, conservation programs in priority watersheds, and groundwater protection and recharge, as well as water resource and infrastructure needs associated with soil and water conservation.[6] The 2014 farm bill also amended and added to existing drought and flood disaster assistance for agricultural producers.[7]

Legislation Considered, but Not Enacted, during the 113th Congress

The 113th Congress considered but did not enact legislation to augment developed water supplies (e.g., water storage, water reuse), settle Indian water rights claims, lift restrictions on firearms at Army Corps projects, and provide direction for individual water projects and facilities. Several bills also would have authorized various regional aquatic ecosystem restoration efforts. These bills generally addressed issues related to the construction of projects for water quality and habitat restoration, as well as governance and reporting for ongoing federal restoration actions.

Several bills related to aquatic ecosystem restoration throughout the country were considered in the 113th Congress and were pending at its conclusion. These bills generally addressed issues related to water quality and habitat restoration, as well as project construction for restoration and water supply allocation among users and the environment. Bills authorizing comprehensive ecosystem restoration initiatives were introduced for the Great Lakes (e.g., H.R. 2773 and S. 1232), Long Island Sound (H.R. 2174 and S. 1080), and Lake Tahoe (H.R. 3390 and S. 1451). These bills addressed governance of ecosystem restoration initiatives and the reauthorization of funding to continue restoration efforts, among other things. Other bills addressed specific aspects of ongoing restoration initiatives, such as expediting restoration project approval in the Everglades ecosystem. H.R. 5764 did pass the House, but was not voted on in the Senate. H.R. 5764 would have authorized the Great Lakes Restoration Initiative and $300 million in appropriations annually from FY2015 to FY2019.

The 113th Congress also considered several different versions of California drought legislation, as well as other legislation to augment water supplies or deal with low water supplies on a broader level. The House twice passed legislation aimed at addressing operation of the federal Central Valley Project (CVP) in California (H.R. 3964 and H.R. 5781) and the Senate also passed legislation to address CVP operations in times of drought emergencies

(S. 2198). However, House and Senate negotiators did not come to agreement on bill language prior to the end of the 113th Congress.

The 113th Congress also considered water research and development legislation targeted at specific programs or issues, such as research related to desalination (e.g., H.R. 745) and state water resources research institutes (S. 970). Similarly, the 113th Congress considered legislation (H.R. 5189 and S. 1971) to coordinate federal research and technologies to better understand and decrease the risks from the interdependencies of the energy (e.g., water use for oil and gas production and power plant cooling) and water sectors (e.g., energy for water transport and treatment).[8]

Unlike the 112th Congress (H.R. 5826), the 113th Congress did not consider legislation addressing the broad federal water science and research portfolio. Water science and research is spread across more than 20 federal agencies. No single water research strategy or formal coordination or prioritization mechanism exists. Some stakeholders are concerned that current research is insufficient to prepare the United States to confront domestic and international water challenges.

Water Resource Considerations for the 114th Congress

The 114th Congress may address some measures left pending at the end of the 113th Congress, and may consider other proposals as well. Because of recent water conditions (e.g., drought in portions of the West and Southwest), disasters, or legal or agency developments, certain basin issues are likely to receive congressional attention. These include the operation of federal reservoirs on the Sacramento and San Joaquin rivers (Central Valley Project in California) and on the Missouri River and its tributaries. Other river basins that may receive attention in the 114th Congress include the Colorado, Klamath, and Rio Grande river basins. Additionally, future operation of Corps facilities on the Columbia River and its tributaries is central to discussions that are underway regarding modification of the Columbia River Treaty with Canada.[9]

Because of recent drought conditions in California and much of the West, Congress might again address drought assistance, planning, and preparedness through oversight hearings and/or legislation, including through Energy and Water Development Appropriations. (See “Drought and Flood Preparedness and Response” section below.)

The 114th Congress may conduct oversight of restoration activities, including those in the Chesapeake Bay, Everglades, Gulf Coast, Great Lakes, San Joaquin River, and Sacramento and San Joaquin Rivers Delta and its confluence with San Francisco Bay (Bay-Delta). Common themes in regional restoration efforts include demand for new project services (e.g., improved or new flood control, water supply, and navigation facilities), protection of threatened and endangered species, drought and flood management, and water quality concerns.

The 114th Congress also may react to efforts by the Administration to implement updated planning guidance for federal water resources projects and to guide federal investment in floodplains. Similarly, Congress may respond to Administration-wide efforts to incorporate climate change adaptation into agency plans and actions, including those being developed by the Corps and Reclamation. The 114th Congress also may engage in discussion of how threatened and endangered species designations and related critical habitat and environmental mitigation requirements affect water resource project construction and operations.

Federal Role in Water Resources

The federal government has long been involved in efforts to facilitate navigation, expand irrigation, and reduce flood and drought losses. For example, nearly every large river basin in the country—from the Columbia, Sacramento, and Colorado rivers in the West to the Missouri, Mississippi, and Delaware rivers—contains one or more federal dam or navigation projects. These projects have largely been constructed by the Corps and Reclamation. More recently, federal involvement has expanded to include municipal water supply development and efforts to protect water-related resources such as fish and wildlife. Increasing pressures on the quality and quantity of available water supplies have resulted in heightened local and regional water use conflicts throughout the country, particularly in the West and Southeast. Pressures include population growth, environmental regulation, in-stream species and ecosystem needs, water source contamination, agricultural and energy water demands, climate change and variability, and changing public interests, such as heightened demand for in-stream recreation.

Congress historically has played a major role in water resources through authorization of and appropriations for regional and site-specific activities; however, numerous responsibilities are split or shared with state, local, and tribal governments, particularly related to water allocation and resource planning and management. Congress also establishes the policies that define the federal role in planning for federal water resource projects, and provides direction for construction, maintenance, inspection, and support of federal projects. Congress makes these decisions within the context of multiple and often conflicting objectives, competing legal decisions, long-established institutional mechanisms (e.g., century-old water rights, and contractual obligations), and in response to events such as floods, droughts, and structural failures.

Federal water resource construction activities shrank during the last decades of the 20th century, marking the end of earlier expansionist policies that had supported large federal investments in dams and hydropower facilities, navigation locks and channels, irrigation diversions, and flood control levees, as well as basin-wide planning and development efforts. Fiscal constraints, changes in national priorities and local needs, few remaining prime construction locations, and environmental and species impacts of construction and operation of federal projects all contributed to this shift. Although these forces are still active, there are proposals for renewed federal financial and technical assistance to address growing pressures on developed water supplies, to manage regional water resources to meet demands of multiple water uses, and to address the aging stock of water resources infrastructure.

Recent drought conditions in the West and Southwest on top of extended and widespread drought in 2012, coastal flooding due to Hurricane Sandy and Hurricane Katrina, and Midwest floods of 2011 have raised other questions about the federal role in water resources. In particular, disasters have brought attention to the trade-offs in approaches to distributing federal appropriations among competing water resources projects, to risk management in water resources, and to the trade-offs in benefits, costs, and risks of the current division of responsibilities among local, state, and federal entities.

FEDERAL WATER RESOURCE AGENCIES

Most of the large dams and water diversion structures in the United States were built by, or with the assistance of, Reclamation or the Corps. Historically, Reclamation projects were designed principally to provide reliable supplies of water for irrigation and some municipal and industrial uses. Corps projects were planned principally to improve navigation and reduce flood damages, with power generation, water supply, and recreation being incidental benefits. Reclamation currently manages hundreds of dams and reservoirs in 17 western states,[10] providing water to approximately 10 million acres of farmland and 31 million people, as well as 58 power plants capable of producing 40 billion kilowatt-hours of electricity annually (enough for approximately 3.5 million homes), and which generate more than $1 billion in revenues annually.[11] The Corps operates nationwide, and its activities are diverse. The Corps has constructed thousands of flood damage reduction and navigation projects throughout the country, including nearly 12,000 miles of commercially active waterways, nearly 1,000 harbors, and 600 dam and reservoir projects (with 75 hydroelectric plants generating 68 billion kilowatt-hours annually). Additionally, the Corps constructed, usually with nonfederal participation, roughly 9,000 miles of the estimated 100,000 miles of the nation's levees, but the agency only maintains 900 miles. The remaining levees are operated by nonfederal entities, often local governments or special districts.

The Natural Resources Conservation Service (NRCS) in the U.S. Department of Agriculture also facilitates water resources development, primarily for flood control in small watersheds and for soil and water conservation purposes.

Many other federal agencies have water-related programs (e.g., the Environmental Protection Agency, the U.S. Geological Survey, the National Oceanographic and Atmospheric Administration, National Aeronautics and Space Administration, Federal Emergency Management Agency, and energy-related agencies such as the Federal Energy Regulatory Commission and Power Marketing Administrations). However, the remainder of this report focuses on the projects, programs, and policies of the Corps and Reclamation.

U.S. Army Corps of Engineers

During most years, the Corps responds to needs arising from flood and drought events, as well as performing its regular activities of constructing and operating and maintaining navigation, flood control, and ecosystem restoration projects and issuing permits for activities that may affect navigable waters and wetlands.[12] As previously noted, Congress authorizes Corps water resources activities and makes changes to the agency's policies generally in an omnibus authorization bill, often titled as a Water Resources Development Act (WRDA). Although WRDA enactment is usually attempted on a biennial schedule, enactment is less regular in part because of multiple and conflicting stakeholder interests and tensions over potential changes in Corps policies. Also, the bill is not a reauthorization bill, per se—rather, it is largely an authorization bill, since few Corps authorities expire.[13] The most recent WRDAs were enacted in 2000, 2007, and 2014. Congress typically appropriates funds for these activities in annual Energy and Water Development Appropriations acts, and, at times, it uses supplemental appropriations bills to fund Corps emergency activities.[14]

Hurricane Sandy in 2012 and Midwest flooding in 2011 raised many questions that the 114th Congress may pursue, including those related to national flood risk and federal actions to reduce that risk. In many cases, Corps facilities and their operations are central to debates over multipurpose river management, especially during drought and flood conditions. For example, reservoir management by the Corps, such as in the Apalachicola-Chattahoochee-Flint basin (which provides much of the water supply for Atlanta, Georgia), often is controversial and has been challenged in the courts. Likewise, Corps operation of dams on the Missouri River and its effect on downstream navigation, flood control, species, and upstream water supplies also remain controversial. Such controversies stem from conflicts over the various authorized uses and purposes for multipurpose projects.

Bureau of Reclamation

Since the early 1900s, Reclamation has constructed and operated many large, multi-purpose water projects, such as Hoover Dam on the Colorado River and Grand Coulee Dam on the Columbia River. Water supplies from these projects have been primarily for irrigation; however, some municipalities also receive water from Reclamation projects. Many of the largest facilities also produce hydropower. Construction authorizations slowed during the 1970s and 1980s due to several factors. In 1987, Reclamation announced a new mission recognizing the agency's transition from a water resources development and construction organization to one primarily occupied with managing water resources, including managing water and related resources in an environmentally and economically sound manner.[15] Since then, increased population, prolonged drought, fiscal constraints, and water demands for fish and wildlife, recreation, and scenic enjoyment have resulted in increased pressure to alter operation of many Reclamation projects. Such changes have been controversial, however, as water rights, contractual obligations, and the potential economic effects of altering project operations complicate any change in water allocation, delivery, or project operations.

In contrast to the Corps, there is no tradition of a regularly scheduled authorization vehicle (e.g., a WRDA) for Reclamation projects. Instead, Reclamation projects are generally considered individually; however, occasionally individual project authorizations are rolled into an omnibus bill, such as P.L. 111-11 enacted in in the 111th Congress or P.L. 102-575 enacted in the 102nd Congress.[16] Because project authorizations are typically enacted in stand-alone legislation, project authorizations and Reclamation bills in general have slowed considerably since the 112th Congress and the onset of congressional earmark moratoria.

As with the Corps, Reclamation river and reservoir management in the face of drought conditions and climate change may also receive congressional attention. In many cases, Reclamation facilities and their operation are central to debates over multi-purpose river management, particularly during times of drought or years of lower than normal precipitation and runoff. For example, controversies associated with Reclamation water resources management in the Sacramento and San Joaquin river watersheds (CA), the Colorado River Basin, and the Klamath River Basin (CA and OR) have often been exacerbated by low water flows and have also been the subject of extended litigation—sometimes even in normal water years. Likewise, ongoing issues associated with Reclamation's operation of pumps in the San Francisco Bay/San Joaquin and Sacramento Rivers Delta (Bay-Delta) and their effect on water users and threatened and endangered species also are quite controversial. This situation also has been exacerbated by low water conditions in some years, including 2014—the third-driest water year on record for California and one of the most extreme drought years on record.[17] Drought and resultant low water supplies are again projected for California and other western areas for the 2015 water year.

Examples of Reclamation-related water project and management issues that may be considered during the 114th Congress include the following:

- response to drought, and operations of federal reservoirs and water delivery;
- regulatory impediments to new water storage projects;
- status of Reclamation's Safety of Dams program;
- authorization, appropriations, and reporting to address aging infrastructure;

- Sacramento-San Joaquin Valley water reliability and species concerns (e.g., Bay-Delta Conservation Plan, CALFED reauthorization, and proposals to address Central Valley Project water supplies);
- miscellaneous project adjustments;
- Klamath River Basin restoration and Klamath project management;
- Colorado River water management;
- San Joaquin River restoration settlement funding and oversight.

A broader issue that could receive attention from Congress is oversight of Reclamation's mission and its future role in western water supply and water resource management generally. As public demands and concerns have changed, so has legislation affecting Reclamation. For example, some project sponsors are considering new partnerships in project development, with project construction largely to be undertaken by nonfederal sponsors. In part, this has developed due to project sponsor frustration in delays over new project studies. Some are pursuing independent nonfederal financing of water resources infrastructure (see section on "Changing Federal Partnerships," below). Further, many in Congress have questioned Reclamation's shift in focus from a water resources *development* agency to a water resources *management* agency and believe Reclamation is not doing enough to develop new water storage. Others argue for increased funds and attention to augment water supplies in the West through water reuse, recycling, aquifer storage and recovery, and desalination technologies. Some also have expressed frustration with regulatory hurdles facing project development and expansions. On the other hand, some groups contend Reclamation has not done enough to protect species and the environment generally.

OVERARCHING POLICY ISSUES

In addition to issues related to federal projects, the 114th Congress faces a number of overarching water resources issues, including flood and drought management and response; project funding and authorization priorities; and aquatic ecosystem restoration.

Drought and Flood Preparedness and Response

Congress is often faced with reacting to natural disasters such as droughts and floods. Drought conditions in California and elsewhere in the West and Southwest, and widespread drought in 2012, have left many areas vulnerable to drought-induced impacts, such as water supply and use limitations, reduced agricultural and power production, and degraded fish and wildlife habitat. Responsibilities for drought planning and response are split among various levels of government and involve many different federal agencies. Although Congress has enacted legislation to coordinate drought information through the National Integrated Drought Information System (NIDIS), there is no overarching national drought policy.

In addition to NIDIS reauthorization (P.L. 113-86) and drought-related provisions of the 2014 farm bill, the 113th Congress enacted legislation (P.L. 113-121) that authorized the Corps to assess its reservoir operations during drought and expanded EPA loan and loan guarantee opportunities and eligibility for water supply systems. Multiple bills in the 113th Congress addressed drought operations of Reclamation facilities (e.g., H.R. 3964, H.R. 4239, and S. 2198). Others addressed water efficiency, conservation, and alternative supplies (e.g., H.R. 5363, S. 2771); several would have facilitated federal or nonfederal water storage projects (e.g., H.R. 3980, H.R. 5412). Additionally, some bills (e.g., S. 2016) proposed changes to the Stafford Act, an emergency assistance act. The majority of these bills consisted of authorizations, with many provisions' implementation contingent upon appropriations; a few bills proposed appropriations to address the western U.S. drought (e.g., H.R. 4039, S. 2016).

Because of ongoing drought conditions in much of the West, Congress might again address drought planning and preparedness through oversight hearings and/or specific legislation.

Periodic but intense flooding also garners attention from Congress. For example, Hurricane Sandy flooding in 2012 and Midwest floods in 2011 tested the nation's emergency response system and resulted in billions of dollars in damages. Although the Corps is the principal flood-fighting agency, other agencies also play a role in flood response and mitigation, such as FEMA's disaster assistance, flood insurance, and pre-disaster mitigation programs. Additionally, responsibilities for flood damage reduction are spread among federal, state, local, and tribal governments. States and local governments in many ways play a primary role in floodplain management because of their jurisdiction over land use decisions and local zoning ordinances— deciding where and how development may occur. The 113th Congress was engaged in some aspects of flood policy: policies affecting FEMA's National Flood Insurance Program,[18] flood damage reduction program and project authorizations in WRRDA 2014, and oversight recovery for areas recently affected by floods. Given the magnitude of the nation's flood risk (e.g., over $10.6 trillion in insured properties in coastal counties on the East Coast and along the Gulf of Mexico)[19] and how the nation's flood risk is increasing,[20] the 114th Congress may consider additional ways to reduce flood risk, such as by improving infrastructure and protecting natural flood mitigation, removing disincentives to improved floodplain management, or promoting more pre-disaster recovery plans for highly vulnerable areas.

Funding and Authorization Priorities

Aging Infrastructure

U.S. water infrastructure is aging; the majority of the nation's dams, locks, and levees are more than 50 years old.[21] Failure of these structures could have significant effects on local communities as well as regional and national impacts. Major capital investments in these structures have been limited in recent years and repairing these facilities would cost billions of dollars.[22] Congressional funding has largely been at the project level and has remained essentially flat, while funding needs have increased over time. To date, no comprehensive reporting or funding solutions to these issues has been enacted. Some propose funding mechanisms that might be more conducive to major capital investments in these projects, such as authorization of loan programs for some infrastructure types, or else including water resource infrastructure among the eligible recipients of funding from an infrastructure bank (such as that proposed in H.R. 2553 in the 113th Congress). Others have proposed utilizing revenues from project beneficiaries (e.g., hydropower revenues, increased user fees) to fund project repairs and upgrades, or even deauthorizing and/or transferring projects to nonfederal entities, such as state or local governments. Still others think that Congress requires more uniform information on the extent of this issue before it considers major funding solutions. In the 113th Congress, the Senate held a hearing on this topic and enacted legislation that would require increased reporting by Reclamation on its aging infrastructure backlog (S. 1800). (See also discussion below on "Changing Federal Partnerships.")

Changing Federal Partnerships

Some have expressed frustration with the pace of authorization for federal water resource projects, and this has resulted in some local sponsors pursuing projects with limited federal partnership or support, or with expectations of future federal reimbursement or credit. An example is the potential construction of Sites Reservoir in California—an off-stream water storage project associated with the federal Central Valley Project (CA). Language authorizing nonfederal construction of proposed federal projects (as long as no federal funding is used) was included in H.R. 1837 and H.R. 6247 in the 112th Congress and H.R. 3964 in the 113th Congress. The FY2014 Consolidated Appropriations Act (P.L. 113-76, H.R. 3547) included a provision authorizing the Secretary of the Interior to partner with local joint power authorities to advance authorized planning and feasibility studies, among other things, including providing grants for such purpose (§208). The 113th Congress (e.g., P.L. 113-121) expanded the ability for nonfederal entities to advance funding for federal projects to spur project construction. Such proposals, however, raise the question of whether federal investment is needed if local sponsors can finance the projects on their own, whether the federal government will be able to meet the expectations for reimbursement, and whether the nonfederal sponsors with available financing will determine which projects get reimbursed from limited federal water resources infrastructure funds.

Another approach was initiated in the 113th Congress through its authorization of Title X of WRRDA 2014, the Water Infrastructure Finance and Innovation Act (WIFIA). The title authorized a pilot program, to be administered by the Corps and the Environmental Protection Agency, for loans and loan guarantees for certain flood damage reduction, public water supply, and wastewater projects. WIFIA was modeled after a similar program that assists transportation projects, the Transportation Infrastructure Finance and Innovation Act, or TIFIA, program.[23]

Earmarks and Project Funding and Authorizations

Water resource project funding is often a part of the debate on congressionally directed spending, or "earmarks." Although water resource project development has historically been directed by Congress, the site-specific nature of the authorizations and appropriations process resulted in projects being subject to earmark disclosure rules and earmark moratoria beginning in the 112th Congress.[24] Earmark moratoria appear to be altering the makeup of Corps and Reclamation appropriations in particular by reducing the congressional additions of specific projects to the budget, and by Congress funding broad categories of activities rather than specific projects. As a result, some projects that have historically benefitted from congressional support have received less (or no) funding in recent enacted appropriations bills.[25] In addition to funding impacts, earmark moratoria have also influenced consideration of site-specific authorizations of water resource projects.

Aquatic Ecosystem Restoration

The 114th Congress may consider the status and priority of major federal efforts to restore aquatic ecosystems that have been altered or impaired by development, habitat loss, and federal water resource projects. Some of these restoration initiatives include those in the Everglades, California Bay-Delta, Great Lakes, Gulf Coast, Chesapeake Bay, Klamath Basin, and elsewhere. The 114th Congress may consider a number of issues pertaining to these ecosystems. For example, Congress may consider legislation to authorize a framework for governance and a comprehensive restoration plan for the Great Lakes and might conduct oversight over the implementation of restoration efforts in the Gulf Coast region. Further, lack of congressional authorization for new construction projects in the Everglades, such as the Central Everglades Planning Project (CEPP), has caused concern that the initiative could be delayed. Congress might consider policies that would streamline authorizations to allow for more projects to be implemented. Funding for existing and new restoration initiatives might generate controversy and could face challenges in the 114th Congress. Deliberations over FY2016 appropriations could also address ecosystem restoration initiatives in various appropriations bills.

End Notes

[1] Earmark moratoria are discussed in the "Overarching Policy Issues" section of this report.

[2] For more information see CRS Report R43298, Water Resources Reform and Development Act of 2014: Comparison of Select Provisions, by Nicole T. Carter et al.

[3] The annual appropriations bill for the Corps and Reclamation is the Energy and Water Development Appropriations bill; however, both agencies occasionally receive funding in emergency or other supplemental appropriations acts, particularly in response to natural disasters, such as floods, droughts, and hurricanes. For more information, see CRS Report R41430, Inland Waterways: Recent Proposals and Issues for Congress, by Charles V. Stern.

[4] In the 113th Congress, the Corps received $5.3 billion in funding related to the landfall of Hurricane Sandy. For more information, CRS Report R42841, Army Corps Supplemental Appropriations: Recent History, Trends, and Policy Issues, by Charles V. Stern and Nicole T. Carter.

[5] See CRS Report IF00014, The 2014 Farm Bill (Agricultural Act of 2014, P.L. 113-79) (In Focus), by Renée Johnson and Jim Monke; and, CRS Report R43076, The 2014 Farm Bill (P.L. 113-79): Summary and Side-by-Side, coordinated by Ralph M. Chite.

[6] For more information on agricultural soil and water conservation programs, see CRS Report R43504, Conservation Provisions in the 2014 Farm Bill (P.L. 113-79), by Megan Stubbs.

[7] For more information on agricultural disaster assistance programs, see CRS Report R43494, Crop Insurance Provisions in the 2014 Farm Bill (P.L. 113-79), by Dennis A. Shields; CRS Report RS21212, Agricultural Disaster Assistance, by Dennis A. Shields; and CRS Report R42854, Emergency Assistance for Agricultural Land Rehabilitation, by Megan Stubbs.

[8] CRS Report R43199, Energy-Water Nexus: The Energy Sector's Water Use, by Nicole T. Carter; CRS Report R43200, Energy-Water Nexus: The Water Sector's Energy Use, by Claudia Copeland.

[9] See CRS Report R43287, Columbia River Treaty Review, by Charles V. Stern.

[10] Arizona, California, Colorado, Idaho, Kansas, Montana, Nebraska, Nevada, New Mexico, North Dakota, Oklahoma, Oregon, South Dakota, Texas, Utah, Washington, and Wyoming.

[11] U.S. Department of the Interior, Budget Justifications and Performance Information Fiscal Year 2014, Bureau of Reclamation, Washington, DC, 2013, p. General Statement-2, http://www.usbr.gov/budget/.

[12] A discussion of the Corps regulatory programs is beyond the scope of this report. The most notable of the Corps' regulatory activities is associated with wetland protection. For more on wetlands issues, see CRS Report RL33483, Wetlands: An Overview of Issues, by Claudia Copeland.

[13] While Corps authorizations generally do not expire or have established sunsets, an automatic de-authorization process begins if projects have not received funding for five years. A number of projects that were authorized in WRDA 2007 (P.L. 110-114) may soon undergo this process.

[14] For more on these topics, see CRS Report R42841, Army Corps Supplemental Appropriations: Recent History, Trends, and Policy Issues, by Charles V. Stern and Nicole T. Carter.

[15] Reclamation's current mission statement can be found at http://www.usbr.gov/main/about/mission.html.

[16] Congress occasionally passes omnibus bills addressing key Reclamation policy changes, as well as new or revised project and program authorizations. Congress enacted P.L. 111-11 in 2009, which included multiple water and land subtitles. The last time Congress enacted a Reclamation omnibus bill was in 1992, the Reclamation Projects Authorization and Adjustment Act (P.L. 102-575).

[17] See http://www.water.ca.gov/waterconditions/.

[18] CRS Report R42850, The National Flood Insurance Program: Status and Remaining Issues for Congress, by Rawle O. King.

[19] AIR Worldwide, The Coastline at Risk: 2013 Update to the Estimated Insured Value of U.S. Coastal Properties, Boston, MA, 2013.

[20] A recent study concluded that the typical 100-year riverine and coastal flood hazard areas are expected to grow nationally by 40% to 45% by 2100. AECOM, The Impact of Climate Change and Population Growth on the National Flood Insurance Program Through 2100, Federal Insurance and Mitigation Administration, Federal Emergency Management Agency, June 2013.

[21] For example, the majority of the Bureau of Reclamation's facilities are more than 50 years old, and Corps infrastructure averages more than 55 years old. See CRS Report RL34466, The Bureau of Reclamation's Aging Infrastructure, by Charles V. Stern.

[22] For example, for the Corps alone, waterway users previously estimated that needed lock repairs and upgrades total $8 billion-$18 billion over the next 20 years, and the Corps has stated that it will require more than $26 billion for dam safety repairs over the next 25 years. Needed repairs to Reclamation facilities totaled $3.2 billion in 2008.

[23] For more information, see CRS Report R43315, Water Infrastructure Financing: The Water Infrastructure Finance and Innovation Act (WIFIA) Program, by Claudia Copeland.

[24] In the 112th Congress, the House Republican Conference, Senate Republican Conference, and the Senate Appropriations Committee all adopted moratoria on earmark requests that are significant to how Congress directs these activities.

[25] See CRS Report R43567, Energy and Water Development: FY2015 Appropriations, coordinated by Mark Holt.

In: Water Resources in the United States ISBN: 978-1-63482-681-5
Editor: Dianna Harrington

Chapter 2

U.S.-MEXICO WATER SHARING: BACKGROUND AND RECENT DEVELOPMENTS*

Nicole T. Carter, Clare Ribando Seelke and Daniel T. Shedd

SUMMARY

The United States and Mexico share the Colorado River and Rio Grande pursuant to binational agreements. Compliance with these agreements becomes more complicated and controversial as water demands near or exceed available supplies and when drought and high heat further reduce availability and increase demand.

Colorado River. The Colorado River flows through seven U.S. states before reaching Mexico, and 97% of the basin is in the United States. A 1944 Water Treaty requires that the United States annually provide Mexico with 1.5 million acre-feet (AF) of Colorado River water, which represents about 10% of the river's average flow. Binational disputes have arisen over water quantity, quality, and conservation. Under the Treaty, disputes can be resolved through amendments, called "minutes." Minute 242 from 1973 requires that the salinity of Colorado River water deliveries not exceed a specified limit. Minute 319, agreed to in 2012,

* This is an edited, reformatted and augmented version of a Congressional Research Service publication R43312, prepared for Members and Committees of Congress, dated January 23, 2015.

provides for a bilateral basin water management, storage, and environmental enhancement effort.

Rio Grande. The Rio Grande is governed by two separate agreements. Deliveries to Mexico in the northwestern portion of the shared basin (near El Paso/Ciudad Juárez) occur under a 1906 Convention, while deliveries for the southeastern portion (which is below Fort Quitman, Texas) are laid out in the 1944 Water Treaty. The 1906 Convention requires an annual delivery to Mexico of 60,000 AF, which can be proportionally reduced based on drought conditions. The United States is not required to make up for reduced deliveries. From 1939 to 2013, deliveries to Mexico were reduced in roughly 30% of the years, including significant reductions in 2012, 2013, and 2014.

Under the 1944 Water Treaty, Mexico has rights to two-thirds of the flows of six Mexican Rio Grande tributaries. The one-third delivered to the United States must average at least 350,000 AF per year, measured in five-year cycles. The United States is entitled to all flows from U.S. tributaries. Mexico met its delivery obligations until the region's 1994-2003 drought.

The current five-year cycle will end in October 2015. Water accounting for the basin has been receiving considerable attention since the second year of the current five-year cycle (October 2011 to October 2012), when Mexico's deliveries were less than 30% of the target. The dry conditions in the basin at that time raised tensions as water constraints intensified. As of late January 2015, Mexico remained behind on its deliveries for the current five-year cycle, although the extent and intensity of the basin's dry conditions had significantly diminished.

The International Boundary and Water Commission (IBWC) is the entity charged with providing binational solutions to issues that arise during application of the water-sharing treaties. Since late 2012, the Mexican and U.S. Sections of the IBWC have been regularly meeting to discuss Mexico's water deliveries and various water accounting concerns.

Legislative Responses. Some Members of Congress have expressed concerns about the adequacy of Mexico's 1944 Water Treaty compliance, U.S. efforts to hold Mexico to its treaty obligations, and the resulting economic impacts, especially in Texas border counties. On December 16, 2014, P.L. 113-235, the Consolidated and Further Continuing Appropriations Act, 2015, became law, and a provision in this omnibus legislation required that the U.S. Section of the IBWC report to the Committees on Appropriations on various water delivery and accounting issues within 45 days of enactment.

U.S.-MEXICO WATER SHARING

How to share water has long been a complex issue for the U.S.-Mexico border region and in the broader U.S.-Mexico relationship.[1] The two countries share a nearly 2,000-mile border. Multiple rivers cross the border or form the border at various points. The principal shared river basins are:

- the Colorado River, which is predominantly in the United States, and crosses the Mexican border on its way to the Gulf of California (*Figure 1*), and
- the Rio Grande, with major tributaries in the United States and Mexico and whose riverbed is the U.S.-Mexico border in Texas (*Figure 2*).

In the 19th century, numerous questions arose regarding the boundaries between the two countries and water sharing of international rivers. Early agreements, starting in 1848, sought to clarify the location of the border.[2] Later in the century, the two countries entered into the Convention of March 1, 1889, establishing the International Boundary Commission (IBC) to apply border agreements.[3] Starting in 1906, agreements to distribute water binationally began to emerge; a 1906 convention on the sharing of Rio Grande for irrigation purposes distributed water in the vicinity of El Paso, Texas.[4] In 1944, the two countries entered into a treaty on "Utilization of Waters of the Colorado and Tijuana Rivers and of the Rio Grande" (hereinafter Treaty or 1944 Water Treaty).[5] The Treaty reconfigured the IBC into the International Boundary and Water Commission (IBWC), which provides binational support and facilitates resolution of issues arising during application of U.S.-Mexico treaties on water quantity, sanitation, water quality, flood control, and boundary demarcation.

This report is a primer on U.S. and Mexican water-sharing topics. It focuses on surface water quantity sharing and recent developments, including drought conditions. Due to Mexico's recent below-target deliveries of Rio Grande water to the United States, particular attention is given to the status, underlying causes, and responses to the Rio Grande water delivery shortfalls. This report describes:

- legal obligations and processes under the 1944 Water Treaty;
- drought conditions from 2010 to 2013;

- water sharing and developments in the Colorado River Basin;
- water sharing in the Rio Grande Basin and Mexico's water delivery shortfalls; and
- stakeholder, diplomatic, and legislative responses to the rate of Mexico's Rio Grande water deliveries.

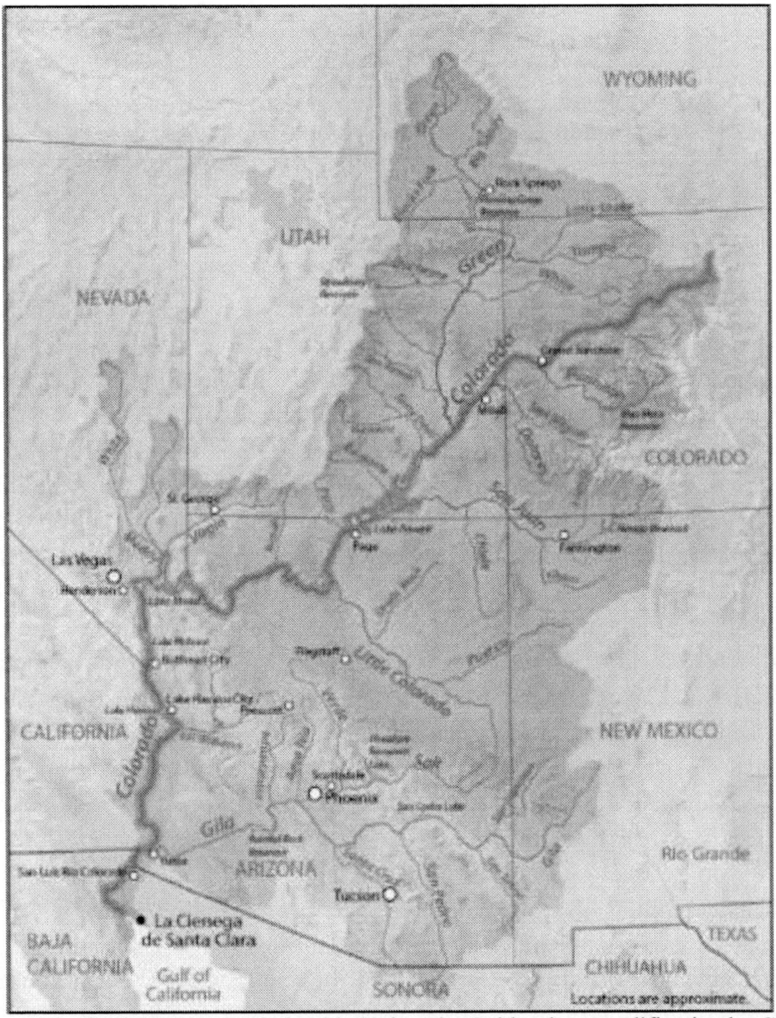

Source: The Earth Institute at Columbia University (with minor modification by CRS), at http://blogs.ei.columbia.edu/wp-content/uploads/2012/12/CO-River-Basin-REVISED.jpg, and modified by CRS.

Figure 1. Illustration of Colorado River Basin.

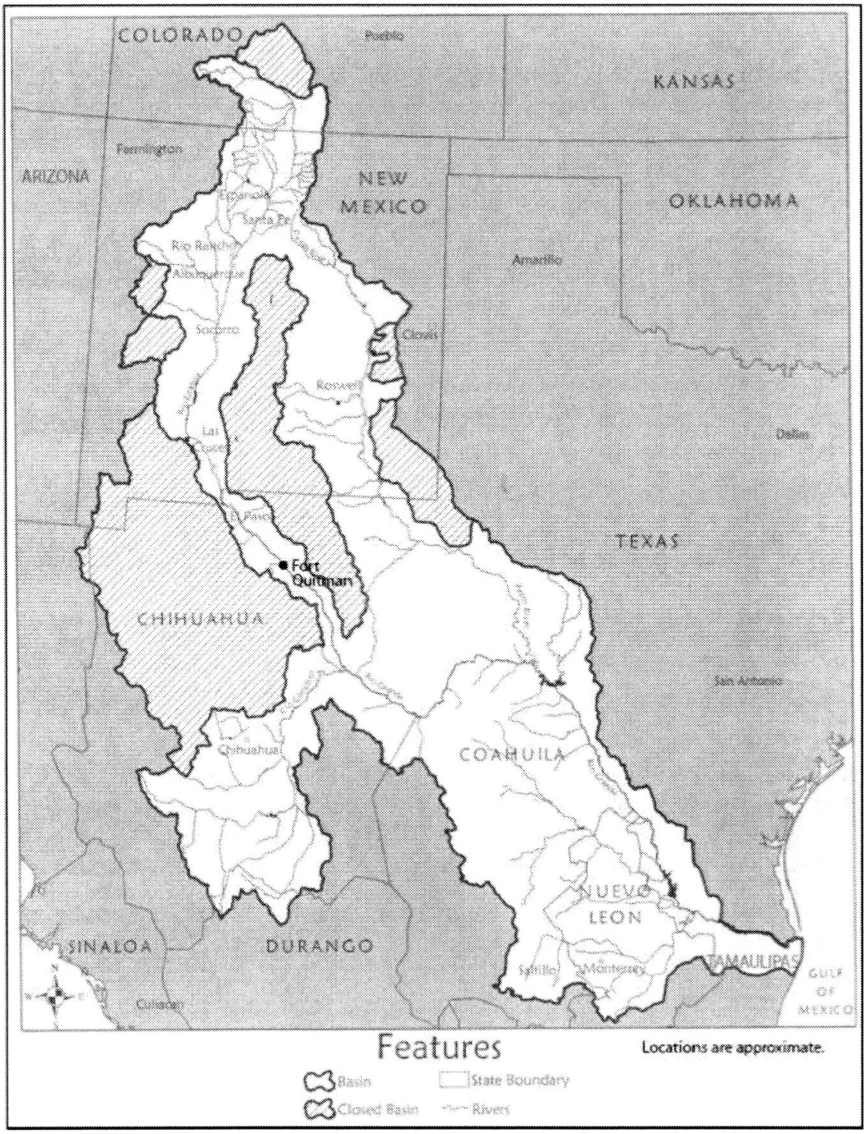

Source: New Mexico Museum of Natural History and Science, at http://www.nmnaturalhistory.org/BEG/ BEG%20Images/MAP_RGB_pg48.jpg, and modified by CRS.

Figure 2. Illustration of Rio Grande Basin.

1944 Water Treaty

Responsibilities, Execution, and Treaty Minutes

The 1944 Water Treaty establishes water allocations for the United States and Mexico and creates the current governance framework for the IBWC to resolve disputes arising from its execution.[6] The IBWC is an international body consisting of U.S. and Mexican Sections, which are overseen by the State Department and the Mexico Ministry of Foreign Relations, respectively.[7] Any disputes that arise under the 1944 Water Treaty are settled through the Treaty's "minute" process. The IBWC is authorized to develop rules and to issue decisions regarding the execution of the Treaty in the form of minutes,[8] which become legally enforceable and essentially amend the Treaty. A proposed minute is forwarded within three days to the government of each country for approval.[9] If the government of either country fails to announce its approval or disapproval within 30 days, the minute is considered approved.[10] If either government disapproves, the matter is removed from IBWC control and the two governments negotiate the issue.[11] If an agreement is reached between the governments, the IBWC must then take any further actions "as may be necessary to carry out such agreement."[12] For the United States, the executive branch has the authority to approve or disapprove of the proposed minutes arising from the Treaty;[13] the President only has the ability to make such agreements pursuant to a treaty if the agreement is in the purview of the treaty.[14]

Water Distribution Requirements

The basic water distribution arrangements in the 1944 Water Treaty are as follows.

- For the Colorado River basin, the United States is to provide Mexico annually with 1.5 million acre-feet (AF)[15] of water.[16]
- For the Rio Grande basin below Fort Quitman, Texas,
  - —Mexico has the rights to two-thirds of the flows that feed into the Rio Grande from the six major tributaries that enter from Mexico: the Conchos, San Diego, San Rodrigo, Escondido, and Salado Rivers and the Las Vacas Arroyo (stream).[17]

- the United States receives all flows from Rio Grande tributaries in the United States and one-third of flows from the six Mexican tributaries.[18] Mexico's water delivery from these six tributaries must average at least 350,000 AF per year, measured in five-year cycles.[19]

If Mexico fails to meet its minimum flow obligations for a five-year cycle because of "extraordinary drought"—a term not defined in the 1944 Water Treaty—it must make up the deficiency during the next five-year cycle with water from the Mexican tributaries.[20] Minute 234 established that Mexico may repay a water debt using three sources of water: (1) excess water from its tributaries; (2) a portion of its allotment from its tributaries; or (3) a transfer of its stored water in the international reservoirs.[21]

If Mexico fails to meet its minimum flow obligations for a five-year cycle and there is no agreement that an extraordinary drought existed, Article 24(d) of the 1944 Water Treaty provides certain mechanisms for dispute settlement. First, the IBWC has the authority "to settle all differences that may arise between the two Governments with respect to ... application of the Treaty."[22] However, if the commissioners are unable to reach agreement on a dispute, the dispute shall be settled through diplomatic channels between the United States and Mexico.[23] Article 24 also provides that the countries may seek recourse in any "general or special agreements which the two Governments have concluded for the settlement of controversies."[24] However, it does not appear that these mechanisms have been necessary in the past. As discussed throughout this report, the United States and Mexico previously have used the minute process and diplomatic efforts to reconcile disputes regarding the 1944 Treaty.

Article 9 of the 1944 Water Treaty provides the IBWC with some flexibility regarding the diversion of water from the Rio Grande. For example, in cases of extraordinary drought occurring in one of the countries, the IBWC may permit water to be withdrawn from the other country in order to help alleviate drought conditions.[25] Further, the IBWC may allow one country to use water allocated to the other country if it can be done "without injury to the latter and can be replaced at some other point on the river."[26] However, if the IBWC authorizes temporary diversions of water from one country to another, the use of such water does not establish a permanent right to divert.[27] Under article 9, the IBWC also is responsible for keeping records concerning the water belonging to both Mexico and the United States.

The 1944 Water Treaty establishes a hierarchy of uses for the water: (1) domestic and municipal uses; (2) agriculture and stock-raising; (3) electric power; (4) other industrial uses; (5) navigation; (6) fishing and hunting; and (7) any other beneficial uses which may be determined by the Commission.[28] A frequent critique of this hierarchy is that it does not include an obligation to maintain water for ecological purposes.[29] In addition, the original 1944 Treaty does not have any provisions that establish requirements for water quality, but only establishes the quantity requirements outlined above.[30] This led to tensions regarding salinity levels in the United States deliveries to Mexico after the 1944 Treaty was ratified.[31] As discussed in the "Salinity" section below, the two countries agreed to Minute 242 in 1973 to resolve the dispute.[32]

Regarding management of reservoirs in the basin that are wholly in one country, the Protocol accompanying the original 1944 Water Treaty establishes that constructed works, such as dams and conveyance structures, in one country that are used only partly for Treaty compliance shall be constructed and operated by the federal agencies of that country, in conformance with the Treaty and in cooperation with IBWC.[33] Subsequent minutes, like Minute 319 (which is discussed below in "Minute 319: Water Conservation and Environmental Protection"), have integrated operational activities in specific circumstances for specific works. To what extent Mexico is operating its reservoirs to support Treaty compliance or prioritizing domestic water demands is a point of debate discussed in the "Stakeholder Perspectives" section below, which focuses on Mexico's Rio Grande water delivery shortfalls.

Other Treaty Provisions

The 1944 Water Treaty established other requirements beyond water distribution obligations. The Treaty, among other things, (1) provided for the construction of certain dams and channels along the rivers;[34] (2) required the IBWC to establish studies and prepare plans for flood control;[35] (3) provided that the IBWC should study and plan for the generation of hydroelectric energy along the rivers;[36] and (4) required the IBWC to establish regulations for the maintenance and operation of reservoirs.[37] Discussion of these treaty requirements is beyond the scope of this report.

Border Aquifers Are Largely Not Addressed by Binational Agreements

Binational aquifers also are shared water resources that can be particularly important for meeting needs during dry times; roughly 20 binational aquifers are significant sources of domestic water supply for overlying populations. For example, the Hueco Bolson aquifer provides Ciudad Juárez's 1.5 million residents and two-fifths of El Paso's 730,000 residents with water. Many border aquifers have experienced significant declines in volume and/or quality. No broad bilateral agreement exists on U.S.-Mexico border groundwater management and use. Declining water levels, deteriorating water quality, and increasing use of groundwater resources have raised concerns about the long-term availability of the border's aquifers. Knowledge about the extent, depletion rates, and quality of transboundary aquifers is limited and in some areas completely absent. A binational aquifer quantity and quality assessment has been initiated, pursuant to the U.S.- Mexico Transboundary Aquifer Assessment Act (P.L. 109-448). The act authorized the Secretary of the Interior, through the U.S. Geological Survey (USGS), to collaborate with the States of Arizona, New Mexico, and Texas through their Water Resources Research Institutes (WRRIs) and with the International Boundary and Water Commission, stakeholders, and Mexican counterparts to provide information and a scientific foundation for state and local officials to address pressing challenges along the U.S.-Mexico border. According to the act's accompanying Senate report (S.Rept. 109-17):

> Ground-water pumping has lowered the water table, depleted aquifers, and reduced the base flow of many streams thus decreasing the quantity of water available to support critical riparian habitats. Excessive ground-water pumping in some major urban centers, such as in the El Paso/Juarez metropolitan region, has caused land subsidence that has damaged homes and essential urban infrastructure. In addition to the effects of ground- and surface-water depletion, degradation of water quality has reduced habitat suitability for the region's diverse biota.

The assessment was authorized in 2006 for $50 million. The USGS used and distributed a total of $2 million through FY2010 for the assessment.

It has received no subsequent funding as of November 2013. Mexico also contributed funding, but estimated funding levels are not available. Additional USGS-supported work is contingent on funding.

Sources: G. E. Eckstein, "Buried Treasure or Buried Hope? The Status of Mexico-U.S. Transboundary Aquifers under International Law," International Community Law Review, vol. 13 (2011), pp. 273-290; W. A. Alley, Five-Year Interim Report of the United States-Mexico Transboundary Aquifer Assessment Program: 2007 -2012, U.S. Geological Survey, Open File Report 2013-1059, Reston, VA, 2013; Christopher E. Wilson, Erik Lee, et al., The State of the Border Report: A Comprehensive Analysis of the U.S.-Mexico Border, Woodrow Wilson Center, El Colegio de la Frontera Norte, Arizona State University, May 2013.

DROUGHT CONDITIONS

Water sharing becomes more complicated during droughts, and both the Colorado River and the Rio Grande basins are prone to multiyear droughts. For the Rio Grande, as shown in *Figure 3,* both 2011 and 2012 were marked by dry conditions, resulting from high heat, low precipitation, and low runoff throughout most of the basin. For the Colorado River, dry conditions developed more noticeably in 2012 and persisted in 2013 and 2014, with dry conditions intensifying in some portions of the basin. Significant rains in late August and September 2013 in some parts of the basin provided some relief, but also resulted in significant flooding and damage, particularly in the state of Colorado.[38]

COLORADO RIVER BASIN

As depicted in *Figure 1*, the Colorado River flows through seven U.S. states (Arizona, California, Colorado, Nevada, New Mexico, Utah, and Wyoming) and forms the border between the Mexican states of Baja California Norte and Sonora, before emptying into the Gulf of California; 97% of the basin is in the United States. Disputes have erupted over the use of the Colorado River water supplies for most of the past century.

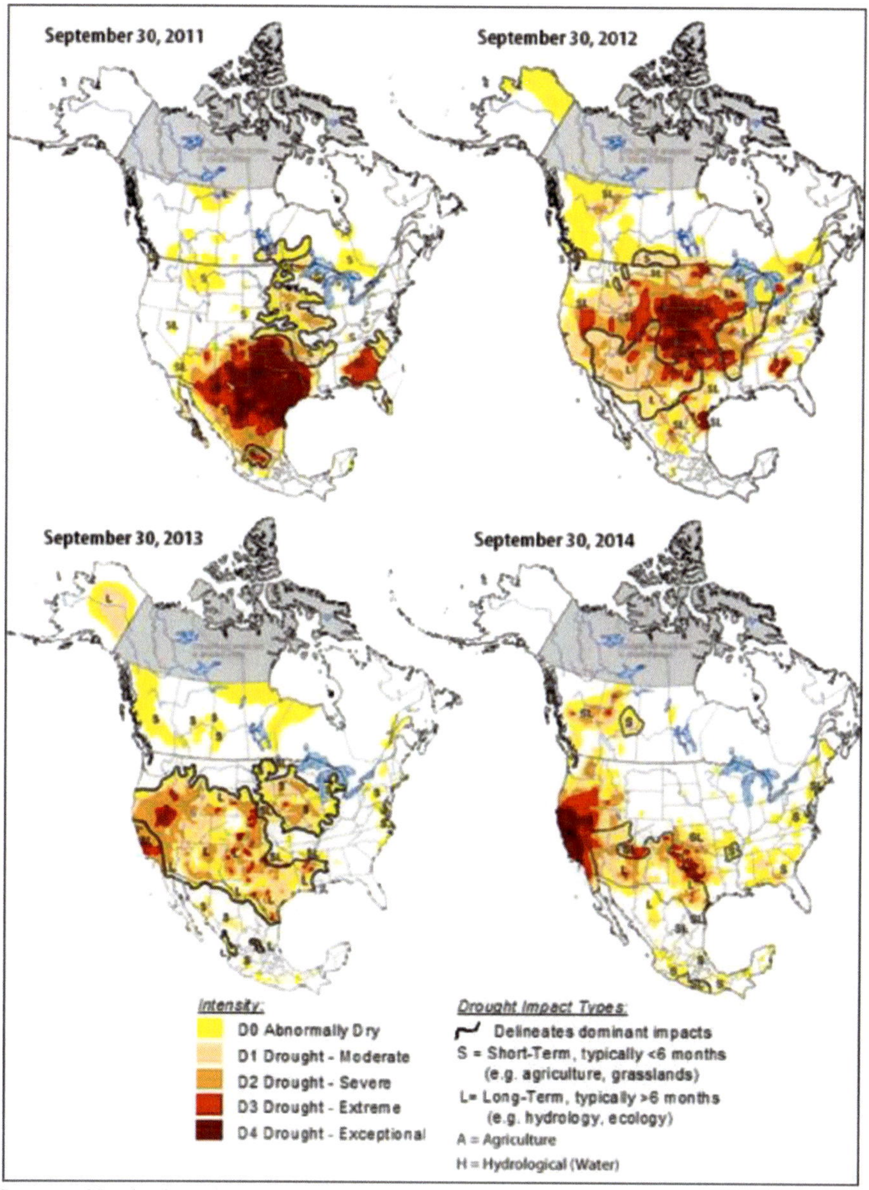

Source: North American Drought Monitor maps (minor modifications by CRS), available at http://www.ncdc.noaa.gov/temp-and-precip/drought/nadm/nadm-maps.php. These maps are created by U.S., Mexican, and Canadian experts who synthesize various drought indices and impacts.

Figure 3. Evolution of North American Drought from 2011 to 2014.

Although many of these disputes have related to state allocations on the U.S. side of the border, issues have also arisen over water quality, availability, and conservation between the United States and Mexico.

When the 1944 Water Treaty was signed, Colorado River flows were estimated at some 16.8 million AF per year. Current flows are now likely closer to 14.4 million AF annually.[39] That is, the 1944 Treaty requirement that the United States provide Mexico with 1.5 million AF annually means that the United States retains roughly 90% of the average flow, but less than originally estimated. In December 2012, the Department of the Interior's Bureau of Reclamation published a study documenting that the demand for the basin's water in the United States in some years exceeds supply, and that the demand-supply imbalance is anticipated to worsen in coming decades.[40] While discussion of Colorado River water issues within the United States is beyond the scope of this report, concern about meeting future demands in the United States is significant to the context of discussions about the basin's water sharing with Mexico.

The following treaty implementation issues in the Colorado River basin are discussed in more detail below: salinity, environmental protection, and Minute 319.

Salinity

While the United States has consistently delivered Mexico's minimum allotment of Colorado River water, disputes did arise about the quality of the water. In the 1960s, salinity in the Colorado River rose dramatically.[41] Mexico was receiving water that was too salty for human, livestock, or agricultural uses. The IBWC helped both countries agree to Minute 218, which took effect in 1965 for a period of five years, requiring the United States to extend a drainage channel to reduce salinity. Five years later, Mexican farmers remained angry about the salinity issue. After the Mexican government threatened to take the water dispute to the International Court of Justice, the United States agreed to Minute 242 in 1973. Per Minute 242, the United States agreed to construct additional channels to control salinity, fund clean-up of the Mexicali Valley lands damaged by the accumulation of salts, and keep salinity levels of delivered water below a certain level. Minute 242 remains in force, and the United States continues to comply with its provisions.

While the IBWC-backed resolution to this crisis proved to be successful, the agreement took a long time and required external pressure to be reached.[42]

Instream Flows for Environmental Protection

The Colorado River Delta at the terminus of the Colorado River, prior to significant expansion of the basin's water consumption, covered 9,650 square miles in the United States and Mexico. The Mexican side of the delta contains wetlands, woodlands and desert areas that are home to many endangered species; part of Mexico's delta is a designated United Nations Biosphere Reserve. According to environmental interests, insufficient water flowing into the delta has contributed to the degradation of 90% of the delta's wetlands;[43] these interests recommend that annual flows accompanied by larger pulses of water every four years would restore the wetlands.[44] These stakeholders have argued that "environmental protection" should be added to the 1944 Treaty as a factor in determining water deliveries to Mexico. Other stakeholders are less supportive of these restoration efforts; some are concerned that they may reduce the allocations available for U.S. users and others do not want to support these efforts while the question of Mexico's compliance with water deliveries in the Rio Grande basin is raising tensions (discussed in "Responses to Mexico's Rio Grande Water Delivery Shortfalls").

The issue of instream flows for environmental protection entered bilateral discussions in the IBWC in the late 1990s. In recent years, bilateral discussions in the basin coalesced around improved management of and conservation of both the Colorado River and its delta. Both governments, along with state officials and conservation groups, worked with the IBWC to develop an agreement that would allocate water to Mexico based on whether there was a surplus or drought and allow for joint investments to create greater environmental protection, as well as greater water conservation (i.e., ability to store water) for Mexico. These discussions led to Minute 319.

Minute 319: Water Conservation and Environmental Protection

Minute 319 was signed on November 20, 2012, and is to be enforced for five years (with the possibility of an extension through 2026 if not supplanted or replaced by another minute). Some view Minute 319 as a step forward in bilateral water management and environmental protection efforts.[45] Others do not support Minute 319 for a variety of reasons, including that the minute signaled increasing cooperation at the same time that water tensions in the Rio Grande basin were particularly acute.

Key elements of the agreement include:

- extending provisions of Minute 318 (Cooperative Measures to Address the Continued Effects of the April 2010 Earthquake in the Mexicali Valley, Baja, California), to allow Mexico to defer delivery of its Colorado River water allocation while Mexico repairs earthquake-damaged infrastructure;
- delivering additional water (i.e., above the 1.5 million AF annual delivery required by the Treaty) to Mexico when water levels are high in Lake Mead;
- reducing deliveries during low Lake Mead reservoir conditions (i.e., Mexico's annual water deliveries would be reduced by up to 0.5 million AF, similar to the reduction by the U.S. lower basin states);[46]
- implementing jointly funded water efficiency and conservation projects to free up water for the Colorado River Delta;
- creating a mechanism by which U.S. water deliveries to Mexico can be held in United States reservoirs for subsequent delivery; and
- continuing to work together to address salinity concerns per Minute 242.

RIO GRANDE BASIN

On most maps, the Rio Grande appears as a continuous line from Colorado to the Gulf of Mexico; in reality, the river dries up at various points. Consequently, what looks like a continuous basin in *Figure 2* actually operates as separate binational basins divided into:

- the northwestern El Paso-Juárez Rio Grande basin from south of Elephant Butte Dam in New Mexico past the water withdrawals and return flows of El Paso, Texas, and Ciudad Juárez, Chihuahua; and
- the southeastern Lower Rio Grande basin, including its tributaries (e.g., Rio Conchos) from Fort Quitman, Texas, to the Gulf of Mexico.

Binational water sharing agreements differ in the two binational Rio Grande basins. In the northwestern El Paso-Juárez basin, the United States is required to deliver water to Mexico. In the southeastern Lower Rio Grande basin, it is largely Mexico that is obligated to deliver water to the United

States. A common characteristic of both basins is that the water demands regularly exceed supply; this imbalance becomes particularly apparent during droughts. While the northwestern El Paso-Juárez water issues have raised significant local concerns recently, the delivery of water from Mexico in the southeastern Rio Grande basin has received the majority of national media and political attention.

Northwestern Rio Grande Basin (El Paso-Ciudad Juárez)

Under the 1906 Convention that guides U.S. deliveries to Mexico at Ciudad Juárez, the United States is to deliver to Mexico 60,000 AF (enough water to irrigate about 25,000 acres) for use in the Juárez Valley of Chihuahua. However, during conditions of extraordinary drought, these deliveries to Mexico are reduced proportionally to reductions in available supplies in the broader basin. From 1939 to 2014, deliveries to Mexico were reduced in roughly 30% of the years; the United States is not required to repay any reduced deliveries. For example, in 2012, U.S. deliveries to Mexico were curtailed, due to drought, to an estimated 23,200 AF (39% of full allotment). In 2013, U.S. deliveries to Mexico were estimated at 3,700 AF (6% of full allotment), and in 2014, the estimated delivery to Mexico was 18,300 (30% of full allotment).

In recent years, U.S. water deliveries to Mexico in the northwestern binational Rio Grande basin have drawn regional attention because the Middle Rio Grande (the portion of the river that traverses New Mexico) has experienced particularly low flow conditions and low storage at reservoirs due to drought. Junior water rights holders (whose water allocations are reduced prior to those with more senior rights) in New Mexico and Texas have received deeply curtailed deliveries (as low as 4% of a full allotment) in recent years.

Specifically, U.S. stakeholders associated with the U.S. Bureau of Reclamation's Rio Grande Project are interested in how water is being delivered to Mexico when the basin is affected by drought. The Rio Grande Project furnishes irrigation water for approximately 178,000 acres in New Mexico and Texas, as well as electric power. Water deliveries from this project have been significantly curtailed as multiple years of dry conditions have depleted reservoir storage.

In particular, the timing of the water releases in 2012 for delivery to Mexico and their potential impacts on U.S. regional interests (e.g., potential conveyance losses because releases for Mexico would not be timed with deliveries to U.S. water districts) raised concerns among some U.S. stakeholders.[47] Mexican growers had sought the surface water deliveries because pumping problems had impaired their ability to start the agricultural season using groundwater.

Southeastern Rio Grande Basin (below Fort Quitman, Texas)

In the southeastern Lower Rio Grande basin, Mexico is required to deliver water to the United States under the 1944 Treaty. As previously noted, the southeastern Lower Rio Grande water delivery account is managed largely on five-year cycles. Mexico's compliance with Treaty delivery requirements often has been accomplished through wet weather flows, rather than purposeful releases from Mexican reservoirs during dry conditions.

Mexico met its deliveries within the five-year cycles until the 1994-2003 drought.[48] During that drought period Mexico accrued a water debt through two five-year water cycles.[49] Diffusion of tensions over the debt was accomplished through presidential intervention, negotiation of new minutes under the 1944 Treaty, and investments in improved water efficiency. Hurricane-induced wet conditions cleared the remaining water debt in 2005.[50] The most significant tributary in the southeastern Rio Grande basin is Mexico's Rio Conchos, which historically contributed 70% of the flow in the Rio Grande, but as of the 1990s was only contributing 40% of the flow. Significant irrigated agricultural production developed in the Rio Conchos basin during the 1980s and early 1990s. It is the change in water deliveries from the Rio Conchos that garnered most of the critical attention during the 1994-2003 drought.

Mexico's Rio Grande Deliveries

The current delivery cycle started October 25, 2010, and is anticipated to end October 24, 2015. In October 2014, at the end of the fourth year of the five-year cycle, Mexico was roughly 339,495 AF (24%) behind in deliveries, based on a total target delivery for those four years of 1,400,000 AF, as shown in *Figure 4.*[51] Although these figures were published by IBWC's U.S Section, adjustments to the agreed-upon water accounting under the treaty may be made subsequent to the release of these delivery figures.

A significant cause of the missed target delivery stems from a deficit of more than 249,000 AF of the annual 350,000 AF target that occurred during the second year of the cycle—that is, deliveries from Mexico were less than

30% of the annual target from October 2011 to October 2012. In the third year of the cycle, Mexico is estimated to have exceeded the target delivery, with roughly 374,000 AF delivered to the United States. In the fourth year of the cycle, available estimates indicate that Mexico delivered less than target annual average allotment.

The timing of the Mexican deliveries within the five-year cycle is a point of tension among some basin interests. This tension was particularly acute during 2012, which falls largely within the second year of the cycle shown in *Figure 4*, as Texas water rights holders faced persistent dry conditions and Mexican water deliveries were below the target deliveries.

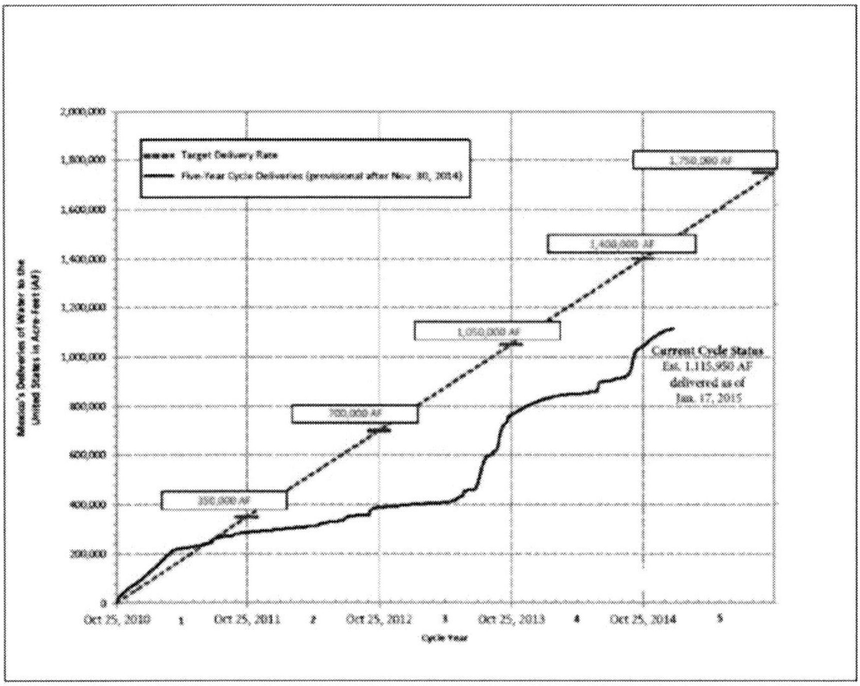

Source: CRS, adapted from IBWC U.S. Section data. Figure shows estimated volumes delivered to the U.S. from six Mexican tributaries named in the 1944 Water Treaty.

Figure 4. Estimated Mexican Deliveries to United States, October 2010 to October 2015 Cycle, Under 1944 Water Treaty.

Two binational reservoirs on the Rio Grande store much of the water Mexico delivers to the United States; these reservoir releases help regulate when the water is delivered to U.S. interests, thereby increasing the value of the delivered water in meeting U.S. water demands. Some U.S. stakeholders argue that the uncertainty regarding the timing of Mexico's deliveries reduces the effective utilization and management of the delivered water, its storage, and its release. Mexico, on the other hand, argues that its deliveries are in compliance with the cycle provided for in the 1944 Water Treaty.

An Over-Allocated Basin

Demands for water in the southeastern Rio Grande basin exceed average supply; it is an over-allocated basin. This imbalance became acute during the 1994-2003 drought. During that drought, the water supply for U.S. agriculture in the Lower Rio Grande basin averaged 78% of the full allocation from 1994 to 1996, and 53% from 1997 to 2004.[52] Currently, Texas water users other than priority water users (i.e., municipal, domestic, and industrial users) can expect to receive on average 70% of their water allocation in average water years.[53]

Over-allocation in Mexico's Rio Grande basin also exists.[54] Much of Mexico's over-allocation is attributed to the expansion of Mexican irrigated agriculture from 1965 to 1994, first in Tamaulipas and later in Chihuahua. Significantly, Texas agricultural water withdrawals did not increase at a similar rate during this period in part because agriculture was already well established on the U.S. side of the border. Growth in industrial activity near the border associated largely with the maquiladora (export assembly plant) industry in Mexico and population growth in the basin's urban areas on both sides of the border also increased demands for urban water supplies.

After the 1994-2003 drought, efforts were made to better align water demand and supply in the southeastern Rio Grande basin; these efforts included buyback of water rights and infrastructure improvements (e.g., reducing water losses from agricultural and municipal water distribution systems). Much of the focus has been on reducing agricultural water use since it accounts for 84% of water withdrawals in the southeastern Rio Grande basin. Some of these efforts were undertaken binationally.[55] Support for some of these investments was provided by Minute 309.[56] Although progress has been made, demand still exceeds supply.[57] Some stakeholders have also questioned how much water savings has been accomplished through these investments and whether the investments in Mexico resulted in improved water deliveries by Mexico under the 1944 Water Treaty.[58]

RESPONSES TO MEXICO'S RIO GRANDE WATER DELIVERY SHORTFALLS

Stakeholder Perspectives

Some U.S. interests contend that Mexico's water delivery process treats U.S. deliveries as a secondary priority to meeting Mexico's own water uses; this has angered some because of the more prescriptive nature of the U.S. water delivery requirement to Mexico (i.e., specified quantities are required to be delivered annually). They point to high storage levels in some Mexican reservoirs as evidence to support their position. They seek the release of waters from these reservoirs to help with the agricultural water needs in the most eastern portion of the basin. These interests see these high reservoir levels as the hoarding of a shared resource.

Other basin stakeholders argue that Mexico's delivery flexibility was explicitly provided in the Treaty to deal with the annual variability of water conditions in the basin. While the flexibility in delivery schedule can be viewed as generous to Mexico, some Mexican interests view the water delivery requirements in the 1944 Water Treaty as generous to the United States. They argue that although 30% of the water in the southeastern Rio Grande basin historically originated in the United States, 50% of the basin's water has been allotted to the United States.[59] This occurs in part because U.S. tributaries are allotted 100% to the United States.

For Mexico, conserving water in its reservoirs can be viewed as part of a long-term drought risk management strategy. The strategy in some Mexican sub-basins to conserve water in some reservoirs during drought also may be influenced by the less developed levels of agricultural insurance and government assistance programs in Mexico. These types of programs in the United States reduce the agriculture sector's economic exposure to droughts and other natural disasters.

Some U.S. stakeholders support reevaluating the current binational water-sharing framework for multiple reasons, including Mexico's Rio Grande water deliveries, its reservoir management and plans, and other disputes and concerns (e.g., environmental restoration and protected and invasive species management issues) in both the Rio Grande and Colorado River basins. Others support continuing to work within the existing Minute and IBWC framework, including some U.S. interests that are encouraged by the resolution and cooperation on binational Colorado River issues and concerned that opening up the 1944 Water Treaty is risky for U.S. interests.

Diplomatic Responses

The IBWC has resolved most border water disputes since 1944, although its processes may be slow to reach resolution. The IBWC employs a combination of technical expertise and diplomacy (backed by the State Department and Mexico's Foreign Ministry) to find solutions that are acceptable to stakeholders on both sides of the border. As with past crises, the IBWC has been the primary entity engaged in resolving the current Rio Grande water dispute over how to address dry conditions and water deliveries in that region.

The U.S. and Mexican Sections of the IBWC have been meeting regularly since late 2012 to discuss Mexico's water deliveries. As of early April 2013, the U.S. Section of the IBWC (USIBWC) reported that the Mexican government had initiated some releases from a reservoir on the San Rodrigo River per the USIBWC's (and the Mexican state of Tamaulipas's)[60] repeated requests.[61] Since April 2013, U.S. and Mexican political officials have stepped in to support IBWC efforts to resolve the current water dispute.

According to U.S. and Mexican officials, the water dispute has been a topic of conversation between high-level government officials, including during President Obama's trip to Mexico.[62] Mexican officials indicate they understand that the United States does not want to wait for the end of this five-year delivery period to receive its allotment of Rio Grande water. The U.S. ambassador to Mexico raised the issue with high-level officials in the administration of Mexican president Enrique Peña Nieto. President Peña Nieto reportedly does not want this dispute to become a serious irritant in the bilateral relationship and has instructed his Foreign Ministry to prioritize working with the IBWC, the State Department, Mexico's Water Commission, and authorities from Texas to reach a mediated settlement to the dispute as soon as possible.[63]

Among the outcomes of diplomatic efforts has been an exchange of technical data to assist in developing options for future water management in the basin. In addition, Mexico delivered more than the 350,000 AF during the third year of the cycle. According to the IBWC Commissioner, "This is a consequence to be sure, of beneficial summer precipitation but also is a direct result of our negotiations achieving a substantially more cooperative approach by Mexican authorities in reservoir management."[64]

Congressional Responses

Several Members of Congress noted the complaints of farmers, local officials, and state officials about the rate of Mexico's water deliveries.[65] Some Members of Congress have expressed concerns about the adequacy of USIBWC and State Department efforts to press Mexico to comply more consistently with its 1944 Water Treaty obligations.

On December 16, 2014, P.L. 113-235, the Consolidated and Further Continuing Appropriations Act of 2015, became law; Section 7045(g)(3) of Division J of the law required the U.S. Section of the IBWC to report to the Committees on Appropriations on various water delivery and accounting issues:

> Not later than 45 days after the enactment of this Act, the Secretary of State, in consultation with the Commissioner for the United States Section of the International Boundary and Water Commission (IBWC), shall report to the Committees on Appropriations on the efforts to work with the Mexico Section of the IBWC and the Government of Mexico to establish mechanisms to improve the transparency of data on, and predictability of, the water deliveries from Mexico to the United States to meet annual water apportionments to the Rio Grande, in accordance with the 1944 Treaty between the United States and Mexico Respecting Utilization of Waters of the Colorado and Tijuana Rivers and of the Rio Grande, and on actions taken to minimize or eliminate the water deficits owed to the United States in the current 5-year cycle by the end of such cycle: Provided, That such report shall include a projection of the balance of the water delivery deficit at the end of the current 5-year cycle, as well as the estimated impact to the United States of a negative delivery balance.

End Notes

[1] For background on the broader U.S.-Mexican relationship, see CRS Report R42917, Mexico: Background and U.S. Relations, by Clare Ribando Seelkehttp://www.crs.gov/pages/Reports.aspx?PRODCODE=R42917.

[2] Treaty of Guadalupe Hidalgo, U.S.-Mex., February 2, 1848, available at http://www.ibwc.state.gov/Files/ Treaty_of_1848.pdf (ending the Mexican-American War).

[3] Convention Between the United States and Mexico regarding the Water Boundary, U.S.-Mex., December 2, 1898, available at http://www.ibwc.state.gov/Files/TREATY_OF_1889.pdf.

[4] Convention of May 21, 1906, on the Equitable Distribution of the Waters of the Rio Grande, available at http://www.ibwc.gov/Files/1906Conv.pdf.

[5] Treaty between the United States of America and Mexico Respecting Utilization of Waters of the Colorado and Tijuana Rivers and of the Rio Grande, U.S.-Mex., Feb. 3, 1944, 59 Stat. 1219 (hereinafter referred to as the 1944 Water Treaty or the Treaty).

[6] Ibid.

[7] Ibid. The members of the IBWC are granted diplomatic status and enjoy "the privileges and immunities appertaining to diplomatic officers" and may "freely carry out their observations, studies and field work in the territory of either country" (ibid.). However, all works and structures that are wholly located within one country—despite the potential international character of such works—remain under the exclusive jurisdiction of the country in which they are located. Each country is responsible for the expenses incurred by their respective section; however, joint expenses are "borne equally by the two Governments." The U.S. Section of the IBWC is typically funded through the annual State Department and Foreign Operations appropriations bill.

[8] Ibid., art. 25.

[9] Ibid.

[10] Ibid.

[11] Ibid.

[12] Ibid.

[13] The Administration has authority to agree to the minutes because they are agreements made pursuant to the 1944 Water Treaty, a treaty that the Senate has ratified. Because a properly enacted treaty is the "Supreme Law of the Land," the power to enter into an agreement required or contemplated by the treaty lies fairly clearly within the President's executive function. U.S. Const. art. VI, § 2 ("the laws of the United States ... [and] all treaties made, or which shall be made, under the authority of the United States, shall be the supreme Law of the Land."); Congressional Research Service, Treaties and Other International Agreements: The Role of the United States Senate, A Study Prepared for the Senate Comm. on Foreign Relations 5 (Comm. Print 2001). For further discussion on the treaty process and executive agreements, see CRS Report RL32528, International Law and Agreements: Their Effect upon U.S. Law, by Michael John Garcia.

[14] Ibid., art. 26.

[15] An acre-foot is about 326,000 gallons of water, enough to cover an acre of land with one foot of water.

[16] Treaty, art. 10.

[17] Ibid., art. 4(A)(c).

[18] Ibid., art. 4(B).

[19] Ibid., art. 4(B)(c).

[20] Ibid., art. 4. For more on compliance, see Allie Alexis Umoff, "An Analysis of the 1944 U.S.-Mexico Water Treaty: Its Past, Present, and Future," Environs: U.C. Davis School of Law Environmental Law and Policy Journal, vol. 32, no. 1 (2008), hereinafter Umoff 2008.

[21] IBWC Minute 234, December 2, 1969.

[22] Treaty, art. 24(d).

[23] Ibid.

[24] Ibid.
[25] Ibid., art. 9(f).
[26] Ibid., art. 9(d).
[27] Ibid., art. 9(e).
[28] Ibid., art. 3.
[29] See, for example, Umoff 2008.
[30] See Treaty.
[31] See Umoff 2008, p. 78.
[32] IBWC Minute 242, Aug 30, 1973.
[33] The Protocol states that for "construction or use of works for storage or conveyance of water, flood control, stream gaging, or for any other purpose, which are situated wholly within the territory of the country of that Section, and which are to be used only partly for the performance of treaty provisions, such jurisdiction shall be exercised, and such functions, including the construction, operation and maintenance of the said works, shall be performed and carried out by the Federal agencies of that country which now or hereafter may be authorized by domestic law to construct, or to operate and maintain, such works. Such functions or jurisdictions shall be exercised in conformity with the provisions of the Treaty and in cooperation with the respective Section of the Commission, to the end that all international obligations and functions may be coordinated and fulfilled." Treaty, Protocol.
[34] Treaty, art. 5.
[35] Ibid., art. 6.
[36] Ibid., art. 7.
[37] Ibid., art. 8.
[38] For more on the causes and consequences of drought, see CRS Report R43407, Drought in the United States: Causes and Current Understanding, by Peter Folger and Betsy A. Cody.
[39] "U.S., Mexico: The Decline of the Colorado River," Stratfor Global Intelligence, May 13, 2013.
[40] U.S. Dept. of the Interior, Bureau of Reclamation, Colorado River Basin Water Supply and Demand Study, December 2012.
[41] Part of the U.S. effort to manage the salinity of its water included the construction of the Yuma Desalting Plant by the Department of the Interior's Bureau of Reclamation. This facility has rarely operated since its construction, however, due in part to the cost of its operations (desalination can require considerable electricity to operate). Instead, the high-saline irrigation water has been disposed (through a canal that enters Mexico and discharges into wetlands called the Ciénega de Santa Clara near the Gulf of California) separately from the United States' required deliveries to Mexico. Whether and how the Yuma Desalting facility should be operated, and how the impacts on the Ciénega de Santa Clara from the reduced discharge of the untreated irrigation runoff should be managed, remain topics of some debate in the basin.
[42] Umoff 2008.
[43] CRS phone interview with Carlos de la Parra, Professor at El Colegio de la Frontera Norte, Mexico, July 8, 2013.
[44] Sierra Club, Regional Conservation Committee: Colorado River Report, February 2001.
[45] CRS phone interview with Carlos de la Parra, Professor at El Colegio de la Frontera Norte, Mexico, July 8, 2013.
[46] Minute 319 stipulates: "the Government of the United States will provide the most current information to Mexico on basin conditions as often as required, including precipitation, stream flow, and water storage conditions in the basin and their historical behavior; the

consumptive water uses for the different basin states, and the historical trend; and the status of the determination of shortage conditions in the Colorado River Basin within the United States ..." This level of data sharing is higher than required in the Rio Grande basin.

[47] Letter from Patrick R. Gordon, Texas Commissioner, Rio Grande Compact Commission, to Edward Drusina, Commissioner, U.S. Section, International Boundary and Water Commission, April 9, 2012; Texas Agrilife Research Center, Drought Watch, February 2013.

[48] During the two five-year cycles between 1992 and 2002, Mexico incurred water debt, failing to deliver the 1,750,000 AF (average annual 350,000 AF) required under the Treaty. See Umoff, 2008. In Minute 293 from October 1995, the United States agreed to loan Mexico water to alleviate the drought. However, in subsequent years Mexico's water debt continued to increase. Minute 308, from June 2002, required Mexico to immediately transfer 90,000 AF of water from international reservoirs to the United States, as partial repayment of the water debt, and required Mexico to conduct studies to improve drought management. After extended negotiations, the two countries reached a solution to eliminate Mexico's water debt for the aforementioned shortages in 2005.

[49] As previously noted, Minute 234, established in 1969, includes a procedure whereby Mexico may pay a water debt using three different sources of water. Minute 234 requires that the deficit payments from these three sources be made concurrently with required deliveries in the following five-year cycle. The United States and Mexico differ in their interpretation and implementation of Minute 234. For example, Mexico claimed that in the event of extraordinary drought, only the deficit incurred during the 1992-1997 five-year water cycle needed to be repaid in the following five-year cycle (i.e., by 2002), and any deficit incurred during the 1997-2002 cycle could be deferred until the next five-year cycle. The United States argued that Minute 234 required that the water debt incurred during the 1997-2002 cycle made up concurrently with the 1992-1997 water debt. The matter was left unresolved.

[50] C. Reed, "The Texas-Mexico Water Dispute and Its Resolution (?): Agricultural Liquid & Land Practice and Discourse along the Rio Conchos, Chihuahua, 1990-2005," (Ph.D. dissertation, The University of Texas at Austin, 2007); hereinafter Reed 2007. For additional information, see footnote 48 of this report.

[51] For the 1992 to 1997 cycle, Mexico's water debt was roughly 1 million AF, which is carried over into the next cycle under the extraordinary drought provision of the 1944 Water Treaty. At the end of the 1997 to 2002 cycle, the debt had grown to roughly 1.5 million AF.

[52] S. Sandoval-Solis, "Water Planning and Management for Large Scale River Basin Case of Study: the Rio Grande/Rio Bravo Transboundary Basin", (Ph.D. dissertation, The University of Texas at Austin, 2011). Hereinafter Sandoval-Solis 2011.

[53] Sandoval-Solis 2011.

[54] Sandoval-Solis 2011.

[55] For example, efforts to improve irrigation efficiency in the largest irrigation district in the Rio Conchos basin were undertaken using assistance from the North American Development Bank (NADBank). NADBank also invested in irrigation efficiency conveyance improvements in U.S. border counties. NADBank provided $40 million in grants for these activities in Mexico, and $40 million for activities in the United States (Reed 2007).

[56] Minute 309, Volumes of Water Saved with the Modernization and Improved Technology Projects for the Irrigation Districts in the Rio Conchos Basin and Measures for Their Conveyance to the Rio Grande, July 3, 2003, available at http://www.ibwc.gov/Files/Minutes/Min309.pdf.

[57] Sandoval-Solis 2011.

In: Water Resources in the United States ISBN: 978-1-63482-681-5
Editor: Dianna Harrington

Chapter 3

DROUGHT IN THE UNITED STATES: CAUSES AND CURRENT UNDERSTANDING*

Peter Folger and Betsy A. Cody

SUMMARY

Drought is a natural hazard, often with significant societal, economic, and environmental consequences. Public policy issues related to drought range from how to identify and measure drought to how best to prepare for, mitigate, and respond to drought impacts, and who should bear associated costs. Severe droughts in 2011 and 2012 in Texas and the midcontinent region, and the current drought in California and the American Southwest, have fueled congressional interest in drought and its near-term effects on water supplies and agriculture, as well as interest in long-term issues, such as drought forecasting and links between drought and human-induced climate change. Continuing drought conditions throughout the country contribute to ongoing interest in drought.

Some part of the country is almost always experiencing drought at some level. Since 2000, no less than 6.6% of the land area of the United States has experienced drought of at least moderate intensity each year. The land area affected by drought of at least moderate intensity varies by year and also within a particular year. For example, since 2000, the total

* This is an edited, reformatted and augmented version of a Congressional Research Service publication R43407, prepared for Members and Committees of Congress, dated February 26, 2014.

U.S. land area affected by drought of at least moderate intensity has varied from as little as 6.6% (July 6, 2010) to as much as 55% (September 25, 2012). Based on weekly estimates of the areal extent of drought conditions since 2000, the average amount of land area across the United States affected by at least moderate-intensity drought has been 27%.

What is drought? Drought has a number of definitions; the simplest may be a deficiency of precipitation over an extended time period, usually a season or more. Higher demand for water for human activities and vegetation in areas of limited water supply, and warmer conditions, increase the severity of drought. For example, drought during the growing season may be considered more severe—in terms of its impacts—than similar conditions when cropland lies fallow.

Some scientists refer to severe drought as a recurring natural disaster in North America. Reconstructions of drought conditions that extend back over 1,000 years—based on observations, historical and instrumental records, and tree-rings—illustrate that portions of the conterminous United States have experienced periods of severe and long-lasting drought termed "megadroughts." Drought reconstructions from tree rings document that severe multidecadal drought occurred in the American Southwest during the 13th century, which anthropologists and archeologists suspect profoundly affected Pueblo society. One looming question is whether the United States, and in particular the American Southwest, including California, may be entering into a period that could be the beginning of a modern-era megadrought. Some scientists postulate that there exists a potential danger of slipping into a new megadrought era akin to the medieval megadroughts if the future climate is warmer due to the buildup of greenhouse gases in the atmosphere. However, regardless of the prospect of human-induced climate change, there is the possibility of return to long-term drought conditions experienced in previous centuries.

Despite advances in scientific understanding of climate and drought, predicting the severity and persistence of severe drought over a specific region of the country is not yet possible more than a few months in advance because of the many factors that influence drought (e.g., precipitation, heat, soil moisture). The prospect of extended droughts and more arid baseline conditions in parts of the United States may challenge existing public policy responses for preparing and responding to drought.

INTRODUCTION

This report discusses how drought is defined (e.g., why drought in one region of the country is different from drought in another region), and why drought occurs in the United States. How droughts are classified, and what is meant by moderate, severe, and extreme drought classifications, are also discussed. The report briefly describes periods of drought in the country's past that equaled or exceeded drought conditions experienced during the 20th century. This is followed by a discussion of the nature and extent of recent droughts that affected Texas and the U.S. midcontinent, and the current drought in California. Lastly, the report discusses future prospects for a climate in the western United States that might be drier than the average 20th-century climate and the possible influence of human-induced climate change.

The likelihood of extended periods of severe drought, similar to conditions experienced centuries ago, and its effects on 21st-century society in the United States raise several issues for Congress. These issues include how to respond to recurrent drought incidents, how to prepare for future drought, and how to coordinate federal agency actions, among other policy choices. A discussion of these issues is beyond the scope of this report.

DROUGHT IN THE UNITED STATES—OVERVIEW

Drought has afflicted portions of North America for thousands of years. Severe, long-lasting droughts may have been a factor in the disintegration of Pueblo society in the Southwest during the 13th century, and in the demise of central and lower Mississippi Valley societies in the 14th through 16th centuries.[1] In the 20th century, droughts in the 1930s (Dust Bowl era) and 1950s were particularly severe and widespread. In 1934, 65% of the contiguous United States was affected by severe to extreme drought,[2] resulting in widespread economic disruption and displacement of populations from the U.S. heartland—many relocating to California's Central Valley—and revealing shortcomings in agricultural and land use practices.

Drought conditions are broadly grouped into five categories: (D0) abnormally dry, (D1) moderate, (D2) severe, (D3) extreme, and (D4) exceptional.[3] Some part of the country is almost always experiencing drought at some level. Since 2000, no less than 6.6% of the land area of the United States has experienced drought of at least moderate intensity each year (*Figure 1*).[4] The land area affected by drought of at least moderate intensity varies by year and also within a particular year. For example, since 2000, the total U.S. land area affected by drought of at least moderate intensity (D1-D4) has varied from as little as 6.6% (July 6, 2010) to as much as 55% (September 25, 2012). Based on weekly estimates of the areal extent of drought conditions since 2000, the average amount of land area across the United States affected by at least moderate-intensity drought has been about 27%.

While the previous percentages refer to the extent of drought nationally, there is particular concern about those locations experiencing the most intense drought conditions: *extreme* and *exceptional* drought. Nearly every year, extreme drought (D3) affects some portion of the country. Since 2000, extreme drought or drier conditions have affected approximately 6.5% of the nation on average.[5] (See *Figure 1.)*

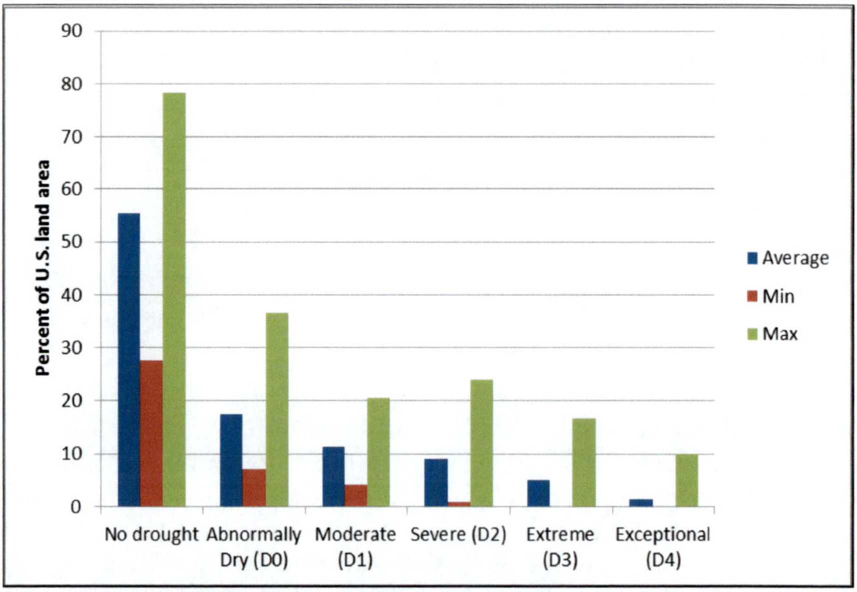

Source: U.S. Drought Monitor, Tabular Data Archive, http://droughtmonitor.unl.edu/MapsAndData/ DataTables.aspx. Modified by CRS.

Notes: Including Alaska, Hawaii, and Puerto Rico.

Figure 1. Average, Minimum, and Maximum Percent of the U.S. Land Area by Drought Category. (January 4, 2000, through February 11, 2014).

Since 2000, exceptional drought (D4) conditions have affected approximately 1.4% of the nation on average. Of particular note were the conditions between June and October 2011; exceptional drought occurred over the largest land area—greater than 9%—during those months, with the affected areas concentrated in Texas. The following year, during August 2012, extreme and exceptional drought extended over 20% of the country and was concentrated in the midsection of the United States.

What Is Drought?

Drought has a number of definitions; the simplest may be a deficiency of precipitation over an extended period of time, usually a season or more.[6] Drought is usually considered relative to some long-term average condition, or balance, between precipitation, evaporation, and transpiration by plants (evaporation and transpiration are typically combined into one term: evapotranspiration).[7] An imbalance could result from a decrease in precipitation, an increase in evapotranspiration (from drier conditions, higher temperatures, higher winds), or both. It is important to distinguish between drought, which has a beginning and an end, and aridity, which is restricted to low rainfall regions and is a relatively permanent feature of an area's climate (e.g., deserts are regions of relatively permanent aridity).[8]

Higher demand for water for human activities and vegetation in areas of limited water supply increases the severity of drought. For example, drought during the growing season would likely be considered more severe—in terms of its impacts—than similar conditions when cropland lies fallow. For policy purposes, drought often becomes an issue when it results in a water supply deficiency: Less water is available than the average amount for irrigation, municipal and industrial supply (M&I), energy production, preservation of endangered species, and other needs. These impacts can occur through multiple mechanisms: decreased precipitation and soil moisture affecting dryland farming; low reservoir levels decreasing allocations for multiple purposes (including irrigation, navigation, energy production, recreation, fish and wildlife needs, and other water supplies); low stream flows limiting withdrawals for multiple purposes, including M&I supplies, among others; decreased exchange of water in lakes resulting in water quality problems limiting recreation (e.g., blue-green algae restrictions in multiple lakes in Oklahoma and Texas during 2011 and 2012 drought conditions). At the national level, drought is monitored and reported by the National Drought Mitigation Center in an index known as the U.S. Drought Monitor, which synthesizes various drought indices and impacts, and represents a consensus view of ongoing drought conditions between academic and federal scientists. Drought can also relate and contribute to other phenomena, such as fires and heat waves.[9]

Drought Classification

To assess and classify the intensity and type of drought, certain measures, or drought indices, are typically used. Drought intensity, in turn, is the trigger for local, state, and federal responses that can lead to the flow of billions of dollars in relief to drought-stricken regions.[10] The classification of drought intensity, such as that shown in *Figure 1*, may depend on a single indicator or several indicators, often combined with expert opinion from the academic, public, and private sectors.

The U.S. Drought Monitor uses five key indicators,[11] together with expert opinion, with indices to account for conditions in the West where snowpack is relatively important, and with other indices used mainly during the growing season.[12] The U.S. Drought Monitor intensity scheme—D0 to D4—is used to depict broad-scale conditions but not necessarily drought circumstances at the local scale. For example, the large regions depicted as red in *Figure 2* faced extreme to exceptional drought conditions for the week of February 11, 2014, but they may contain local areas and individual communities that experienced less (or more) severe drought.

Drought Is Relative

Drought and "normal" conditions can vary considerably from region to region. For example, the U.S. Drought Monitor shows that the San Francisco Bay Area faced extreme drought in mid-February 2014. Similarly, the city of Reno, Nevada, was in extreme drought during the same time period. However, Reno receives on average a total of 2.36 inches of precipitation over the three-month period prior to February: November through January.[13] In contrast, San Francisco receives on average 10.6 inches of precipitation over the same period.[14] Both cities faced extreme drought in early February 2014, but what was normal for San Francisco was very different from what was normal for Reno.[15]

To deal with these differences, meteorologists use the term meteorological drought—usually defined as the degree of dryness relative to some average amount of dryness and relative to the duration of the dry period. Meteorological drought is region-specific because atmospheric conditions creating precipitation deficiencies vary from region to region, as described above for Reno and San Francisco.

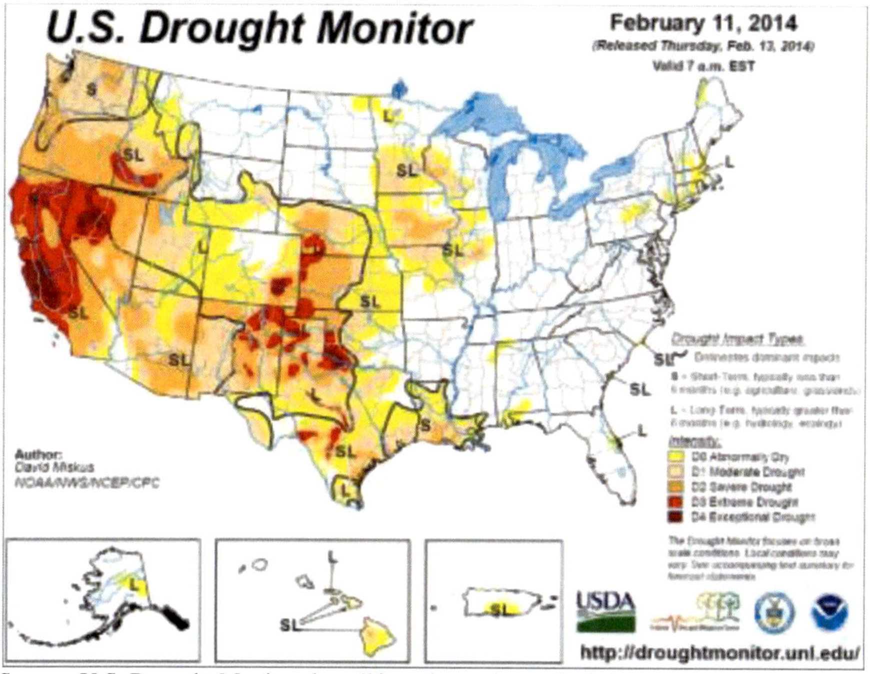

Source: U.S. Drought Monitor, http://droughtmonitor.unl.edu/.
Notes: The U.S. Drought Monitor national maps are updated weekly.

Figure 2. Example of a U.S. Drought Monitor Map.

Drought Is Multifaceted

In addition to the color-coded D0-D4 designations, U.S. Drought Monitor maps often include an "S" and "L" designation to provide additional information about the nature of drought. (See *Figure 2*.) The "S" designation is intended to indicate a combination of drought indices that reflect impacts that respond to precipitation over several days up to a few months (short-term effects). These would include impact to agriculture, topsoil moisture, unregulated streamflows, and aspects of wildfire danger. The "L" designation approximates responses to precipitation over several months up to a few years (long-term effects). These would include reservoir levels, groundwater, and lake levels. As *Figure 2* shows, some regions of the United States include both an "S" and "L" designation, indicating that in mid-February 2014 those regions are experiencing both short and long-term impacts.

WHAT CAUSES DROUGHT IN THE UNITED STATES?

The immediate cause of drought is:

> the predominant sinking motion of air (subsidence) that results in compressional warming or high pressure, which inhibits cloud formation and results in lower relative humidity and less precipitation. Regions under the influence of semi permanent high pressure during all or a major portion of the year are usually deserts, such as the Sahara and Kalahari deserts of Africa and the Gobi Desert of Asia.[16]

Prolonged droughts occur when these atmospheric conditions persist for months or years over a certain region that typically does not experience such conditions for a prolonged period.[17]

Predicting drought is difficult because the ability to forecast surface temperature and precipitation depends on a number of key variables, such as air-sea interactions, topography, soil moisture, land surface processes, and other weather system dynamics.[18] Scientists seek to understand how all these variables interact and to further the ability to predict sustained and severe droughts beyond a season or two, which is the limit of drought forecasting abilities today.

In the tropics, a major portion of the atmospheric variability over months or years seems to be associated with variations in sea surface temperatures (SSTs). Since the mid- to late 1990s, scientists have increasingly linked drought in the United States to SSTs in the tropical Pacific Ocean. Cooler than average SSTs in the eastern tropical Pacific region—"La Niña-like" conditions—have been shown to be correlated with persistently strong drought conditions over parts of the country, particularly the West.[19] A number of studies have made the connection between cooler SSTs in the eastern Pacific and the 1998-2004 western drought,[20] three widespread and persistent droughts of the late 19th century,[21] and past North American "megadroughts" that occurred between approximately 900 and 1300 A.D.[22] The precolonial megadroughts apparently lasted longer and were more extreme than any U.S. droughts since 1850, when instrumental records began. Some modeling studies suggest that within a few decades the western United States may again face higher base levels of dryness, or aridity, akin to the 900- 1300 A.D. period.[23]

Although the relationship between cooler than normal eastern tropical Pacific SSTs (La Niña-like conditions) and drought is becoming more firmly established, meteorological drought is probably never the result of a single cause. Climate is inherently variable, and accurately predicting drought for one region in the United States for more than a few months or seasons in advance is not yet possible because so many factors influence regional drought. What is emerging from the scientific study of drought is an improved understanding of global linkages—called teleconnections by scientists—between interacting weather systems, such as the El NiñoSouthern Oscillation, or ENSO. (See box for a description of ENSO.) For example, some scientists link La Niña conditions between 1998 and 2002 with the occurrence of near-simultaneous drought in the southern United States, Southern Europe, and Southwest Asia.[24]

El Niño-Southern Oscillation (ENSO)

Under normal conditions, the trade winds blow toward the west in the tropical Pacific Ocean, piling up the warm surface waters so that the ocean surface off Indonesia is one-half meter higher than the ocean off Ecuador. As a result, deep and cold water flows up to the surface (upwelling) off the west coast of South America. The upwelling waters are 8 degrees Celsius (14.4 degrees Fahrenheit) cooler than waters in the western Pacific. During El Niño, the trade winds relax, upwelling off South America weakens, and sea surface temperatures rise. The El Niño events occur irregularly at intervals of 2-7 years, and typically last 12-18 months. These events often occur with changes in the Southern Oscillation, a see-saw of atmospheric pressure measured at sea level between the western Pacific and Indian Ocean, and the eastern Pacific. Under normal conditions, atmospheric pressure at sea level is high in the eastern Pacific, and low in the western Pacific and Indian Oceans. As implied by its name, the atmospheric pressure oscillates, or see-saws, between east and west; and during El Niño the atmospheric pressure builds up to abnormally high levels in the western tropical Pacific and Indian Oceans—the El Niño-Southern Oscillation, or ENSO. During a La Niña, the situation is reversed: Abnormally high pressure builds up over the eastern Pacific, the trade winds are abnormally strong, and cooler-than-normal sea surface temperatures occur off tropical South America. La Niña-like conditions have been shown to be correlated with persistently strong drought conditions over parts of the country, particularly the West. Scientists use the terms ENSO or ENSO cycle to include the full range of variability observed, including both El Niño and La Niña events.

It is widely recognized that ENSO drives substantial variability in rainfall and severe weather, including drought. Some recent studies indicate that human-forced changes in climate may alter the spatial pattern of ENSO-driven variability in precipitation and temperature, for example increasing El-Nino drying in the western Pacific Ocean, and increasing rainfall in the central and eastern equatorial Pacific. Other studies, including the 2007 IPCC report, point to a lack of consensus over global-warming driven changes in the timing and intensity of El-Nino driven events, and suggest that it is not yet possible to say that ENSO activity will be enhanced or damped, and whether the frequency of ENSO events will change. Some scientists share the opinion that the instrumental record is too short to document the spectrum of ENSO variability and there is little knowledge of how variability alters with changes in the climate.

Source: Tropical Ocean Atmosphere Project, Pacific Marine Environmental Laboratory, at http://www.pmel.noaa.gov/tao/proj_over/ensodefs.html; Scott Power et al., "Robust twenty-first-century projections of El Niño and related precipitation variability," Nature, published online October 13, 2013; Meehl, G. A. et al., in Climate Change 2007: The Physical Science Basis. Contribution of Working Group I to the 4th Assessment Report of the Intergovernmental Panel on Climate Change, eds Solomon, S. et al., Cambridge Univ. Press, 2007; Mat Collins et al., "The impact of global warming on the tropical Pacific Ocean and El Niño," Nature Geoscience, published online May 23, 2010, Pascale Braconnot et al., "Editorial: El Niño-Southern Oscillation-observations and modeling," PAGES news, vol. 21, no. 2, August 2013, p. 48.

Prehistorical and Historical Droughts in the United States

Some scientists refer to severe drought as "the greatest recurring natural disaster to strike North America."[25] That claim stems from a reconstruction of drought conditions that extends back over 1,000 years, based on observations, historical and instrumental records where available, and on tree-ring records or other proxies in the absence of direct measurements.[26] What these reconstructions illustrate is that the conterminous United States has experienced periods of severe and long-lasting drought in the western states and also in the more humid East and Mississippi Valley. The drought reconstructions from tree rings document that severe multidecadal drought occurred in the American Southwest during the 13th century, which anthropologists and archeologists suspect profoundly affected Pueblo society.

Tree ring drought reconstructions also document severe drought during the 14th, 15th, and 16th centuries in the central and lower Mississippi Valley, possibly contributing to the disintegration of societies in that region.[27]

More recently, a combination of tree ring reconstructions and other proxy data, historical accounts, and some early instrumental records identify three periods of severe drought in the 19th century: 1856-1865 (the "Civil War drought"), 1870-1877, and 1890-1896.[28] The 1856-1865 drought, centered on the Great Plains and Southwest, was the most severe drought to strike the region over the last two centuries, according to one study.[29] The 1890-1896 drought coincided with a period in U.S. history of federal encouragement of large-scale efforts to irrigate the relatively arid western states under authority of the Carey Act.[30] Congressional debate also occurred over a much larger federal role in western states irrigation, which led to the Reclamation Act of 1902.[31]

In the 20th century, the 1930s "Dust Bowl" drought and the 1950s Southwest drought are commonly cited as the two most severe multiyear droughts in the United States.[32] (The 1987- 1989 drought was also widespread and severe, mainly affecting the Great Plains but also instigating extensive western forest fires, including the widespread Yellowstone fire of 1988.) According to several studies, however, the 19th and 20th century severe droughts occurred during a regime of relatively less arid conditions compared to the average aridity in the American West during the 900 to 1300 A.D. megadroughts. One study indicates that the drought record from 900 to 1300 A.D. shows similar variability—drought periods followed by wetter periods—compared to today, but the average climate conditions were much drier and led to more severe droughts.[33]

RECENT DROUGHTS: TEXAS, THE MIDCONTINENT AND CALIFORNIA

Over the past five years, portions of the country have been gripped with extensive and extreme to exceptional drought. As noted above, drought conditions nearly always occur in some part of the United States; however, the intensity of the Texas drought in 2011-2013, the widespread nature of the midcontinent drought in 2012-2013, and the return of drought to California in 2013-2014 have focused national and congressional attention on those regions. The national importance of agriculture in those regions combined with drought conditions (i.e., lack of precipitation combined with demand from farming and ranching) illustrates the nature of drought: namely, it is the combination of lack of precipitation and acute demand for water. The following section briefly describes these specific regional droughts.

Drought in Texas: 2011-2013

In mid-February 2011, nearly 90% of Texas was experiencing abnormally dry to drought conditions, and nearly 8% of the state was in extreme drought (D3).[34] The 2011 drought in Texas represented a dramatic shift compared to the same time period in 2010, when approximately 7% of the total land area in Texas was abnormally dry, without even moderate drought in any part of the state.[35] (See *Figure 3*, comparing 2010, 2011, 2012, 2013, and 2014.)

Drought conditions worsened in Texas through the beginning of October 2011, when 88% of the state experienced exceptional drought conditions (and only 3% of the state was not classified as extreme or exceptional drought).[36] Drought conditions generally improved throughout the rest of 2011, but large portions of the state were still affected by extreme or exceptional drought until late winter and early spring of 2012, when the eastern portion of the state recovered to normal or abnormally dry conditions (the least severe category) because of above-normal rainfall from December 2011 through February 2012.[37] Although drought conditions persisted in parts of Texas up through early 2014, the most severe, extreme, and exceptional conditions occurred in 2011.

According to Texas state climatologist John Nielsen-Gammon, 2011 may have been the worst one-year drought on record for Texas.[38] Compounding the effects of abnormally low precipitation, the June-August average temperature in Texas was approximately 2.5 degrees Fahrenheit (F) greater than any previous Texas summer since 1895 and 5 degrees F greater than the long-term average.[39]

Drought in the Midcontinent: 2012-2013

In mid-August 2012, approximately 70% of the land area of the United States (including Alaska and Hawaii) was affected by abnormally dry and drought conditions.[40] The land area affected by abnormally dry or drought conditions stayed at or above 65% through February 2013.

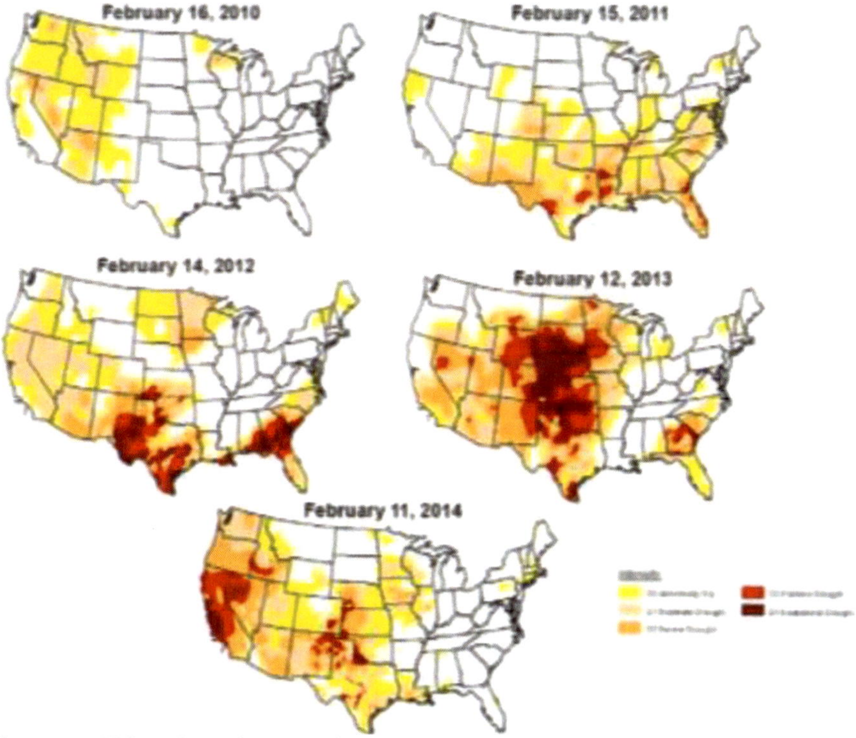

Source: U.S. Drought Monitor, http://droughtmonitor.unl.edu/MapsAndData/MapArchive.aspx.

Note: Alaska and Hawaii not shown.

Figure 3. Drought Conditions in the United States: 2010-2014.

The intensity of the drought varied, with the regions of extreme and exceptional drought were clustered across the Midwest, Great Plains, Southwest, and in the Southeast, particularly Georgia in 2012. (See *Figure 3.)*

Figure 3 shows that Texas and portions of Florida and Georgia experienced exceptional drought conditions (the worst category of drought) in early 2012, while the upper Midwest, including most of the Mississippi Valley experienced normal conditions. A year later, in early 2013, the drought had eased somewhat in portions of Georgia and Florida, but intensified throughout the center of the country from Texas to the Canadian border. Nearly 12% of the contiguous United States was in exceptional drought conditions from late June 2011 through October 2011, compared to approximately 6% of the country the following year.[41] However, exceptional drought conditions persisted over nearly 6% of the contiguous United States from mid-August 2012 through mid-

February 2013. Although less severe for portions of the country, such as Texas and Florida, the 2012-2013 drought affected broader swaths of the agricultural heartland compared to 2011 (*Figure 3*). The 2012-2013 experience illustrates that the extent, timing, and particular features of areas affected by drought—dryland versus irrigated farm regions, regions that are still recovering from previous droughts, or regions with multi-year surface storage or ample groundwater resources—are important in addition to the relative severity of drought conditions.

Origin of the 2012-2013 Drought

Figure 3 shows a snapshot of drought conditions for mid-February in the United States for 2010- 2014. In 2010 most of the United States was experiencing near-normal conditions. The extent and severity of the 2012-2013 drought raised questions regarding its origin, and whether the drought was within the range of natural variability in the U.S. Midwest and Plains, or whether it was linked to longer-term changes in the Earth's climate system, such as human-induced global warming.

Although the images presented in *Figure 3* may seem to indicate a steady progression of drought in the middle portion of the country from near-normal conditions in 2010 to widespread and intense drought in 2012-2013, a March 2013 analysis concludes that the 2012 intense drought in the midcontinent region was a discrete extreme event.[42] The report states that "the event did not appear to be just a progression or a continuation of the prior year's record drought event that developed in situ over the central U.S."[43] Instead, the report asserted that the drought developed suddenly, with near normal precipitation during winter and spring 2012 over the Great Plains. The drought resulted from an extreme lack of precipitation during the summer months: 2012 was the driest summer in the historical record for the region, experiencing even less rainfall than the years 1934 and 1936, when the central Great Plains were about 0.5° C warmer than 2012.[44] Essentially the rains abruptly stopped in May over the central Great Plains, and did not return for the summer.

The report further stated that the 2012 summer drought was a "climate surprise," because summertime Great Plains rainfall has been trending upward since the early 20th century, and the last major drought occurred in 1988.[45] Further, the report concluded that neither sea surface temperatures, which have been rising generally due to global warming, nor changes in greenhouse gases in the atmosphere, were responsible for producing the anomalously dry conditions over the central Great Plains in 2012.[46] (See section on "Drought and Climate Change," below, for a brief discussion of climate change and drought.)

Change Over a Year: Drought in California, 2013-2014

By early 2014 extreme and exceptional drought conditions had fully developed in California and parts of the Pacific Northwest, while the Midcontinent drought of 2012-2013 lessened somewhat. (See *Figure 3*.) The percentage of the conterminous United States experiencing drought of any category—D0 through D4—shrunk between February 2013 and February 2014 (*Table 1*).

Table 1. Percentage of Area Affected By Drought—Conterminous United States and State of California—Comparison Between 2013 and 2014

Week	Region	No drought	D0-D4	D1-D4	D2-D4	D3-D4	D4
2/11/14	U.S.	44.54	55.46	37.60	21.41	6.86	0.88
2/12/13	U.S.	34.11	65.89	55.73	37.74	17.71	6.61
2/11/14	California	1.43	98.57	94.54	91.59	60.94	9.81
2/12/13	California	34.53	65.47	47.18	23.72	0	0

Source: U.S. Drought Monitor, http://droughtmonitor.unl.edu/MapsAndData/ Weekly Comparison.aspx.

Although a smaller percentage of the U.S. land mass is experiencing drought in mid-February 2014 compared to a year earlier, the percentage of California undergoing drought has increased markedly over the same time period. *Table 1* shows that while 65% of the state experienced abnormally dry and drought conditions in February 2013, no portion of the state was in extreme or exceptional drought.

In contrast, a year later over 98% of the state was abnormally dry or in drought and nearly 61% of the state was facing extreme to exceptional drought, including portions of the Central Valley, one of the richest agricultural regions in the United States. (See *Error! Reference source not found.*.) Moreover, as of February 2014 snowpack in the Sierra Mountains was well below normal, and water levels in multi-year reservoirs were below average conditions for that time of year. [47] These conditions have resulted in record cut-backs to state and federal water supply contractors.

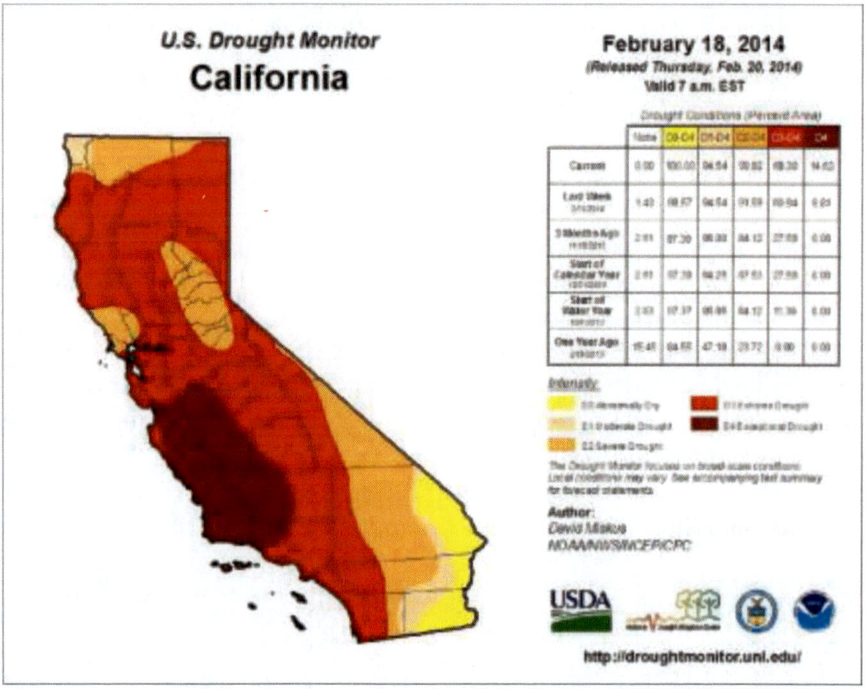

Source: U.S. Drought Monitor, http://droughtmonitor.unl.edu/Home/StateDrought Monitor.aspx?CA.

Figure 4. California Drought Conditions.

Although California experienced the driest year on record in 2013, the severity of drought impacts is not solely related to the lack of precipitation and the duration of dryness. For example, drought risks to U.S. citizens generally increase in proportion to the population affected by drought, as well as in proportion to the resources and economic activity dependent upon average water supply and soil conditions. This situation is illustrated currently in California where more than 22 million people rely on water delivered from the Sacramento and San Joaquin Rivers Delta confluence with San Francisco Bay (Bay-Delta) – also the home and nursery for several species of federal and state listed threatened and endangered species. Farms in the state's Central Valley also rely on water coming from the Bay-Delta, as well as run-off from the Sierra mountains. With snowpack and reservoirs both at record lows, and California being the top U.S. agricultural producer in terms of cash receipts, drought impacts in California could be felt nationwide.[48]

The 2007-2009 California Drought: Setting the Stage for the 2014 Drought?

The 2007-2009 California drought[49] was complicated by decades of tension over water supply deliveries for irrigation and municipal and industrial uses, and the preservation of water flows to protect threatened and endangered species. Dry conditions that began in 2007 continued through the 2009 water year (October 2008 through September 2009) and into the fall of 2009. According to the California Department of Water Resources, the 2007-2009 drought was the 12th-driest three-year period in California history since measurements began.[50] Although hydrological conditions were classified as below normal in 2010 and "wet" (well above average) in 2011, the 2012 water year was classified as "below normal" for the Sacramento River basin and "dry" for the San Joaquin River basin.[51] Above-average reservoir storage at the end of 2011 mitigated reductions to water users. Although the drought was declared over[52] in spring 2011, by August 2012, the U.S. drought monitor again showed increasing severity of drought in the eastern portion of the state. Water deliveries to state and federal water project contractors were restricted again in 2012, as well as in 2013.[53]

California's dry conditions from 2007 through 2009 exacerbated an already tight water supply, where federal and state water deliveries had been reduced in response to a court order to prevent extinction of the Delta smelt.[54] These factors are still in play in 2014 and are challenging water managers charged with protecting certain species from extinction while also supplying water for farms, cities, and fish and wildlife.

DROUGHT AND CLIMATE CHANGE

The relationship between climate change and future trends in droughts is complex and its scientific understanding appears to be evolving. In 2007 the Intergovernmental Panel on Climate Change (IPCC) released its Fourth Assessment Report, which stated that, globally, very dry areas have more than doubled since the 1970s due to a combination of El Nino-Southern Oscillation (ENSO)[55] events and global surface warming.[56] The 2007 IPCC report added that very wet areas declined by about 5% globally. The report asserted that documented trends in severe droughts and heavy rains show that hydrological conditions are becoming more intense in some regions.[57]

In 2012, the IPCC issued a new report that noted: "There are still large uncertainties regarding observed global-scale trends in droughts."[58] The new report noted that its earlier assessment, that very dry areas have more than doubled since the 1970s, was based largely on only one study, which relied on a measurement largely related to temperature, not moisture. A different study, which looked at soil moisture simulations, found that global trends in drought duration, intensity, and severity predominantly were *decreasing*, not increasing, but with strong regional variation.[59]

The 2012 IPCC report assigned medium confidence[60] that there has been an overall slight tendency toward *less* dryness in North America (i.e., a wetting trend with increasing soil moisture and runoff). It noted that the most severe droughts in the 20th century occurred in the 1930s and 1950s, where the 1930s drought was the most intense and the 1950s drought was the most persistent. In comparison to the severe megadroughts that occurred in North America hundreds and thousands of years ago, as documented using paleoclimate evidence (discussed earlier in this report), these recent droughts were not unprecedented, according to the 2012 IPCC report.

The 2012 report concluded that despite new studies that have furthered the understanding of mechanisms leading to drought, there is still limited evidence to attribute observed changes. The IPCC assessed that there was medium confidence that anthropogenic influence has contributed to changes in drought patterns in the second half of the 20th century, but gave low confidence to the attribution of changes in drought patterns at the regional level.[61] The report noted that some regions of the world have experienced trends towards more intense and longer droughts, such as southern Europe and West Africa. But in other regions, such as central North America and northwestern Australia, droughts have become less frequent, less intense, or shorter. How the 2011-2012 drought in the central United States and the current California drought may change that assessment in the forthcoming IPCC report on impacts, adaptation, and vulnerability (expected later in 2014) remains to be seen.[62]

Further adding to the complexity and challenge to the scientific understanding of what causes drought was work presented at the European Geosciences Union meeting in April 2013 that attempted to simulate megadroughts that occurred in the past.[63] The simulations produced a number of megadroughts that lasted for decades; however, they did not match the timing of the past documented megadroughts. The scientists presenting their work at the meeting concluded that the model they used seemed to miss some of the dynamics that drive large droughts.

Despite the current lack of predictive capability for forecasting megadroughts, work presented at the American Geophysical Union meeting in December 2013 underscored the finding that in the past the climate system has the inherent capacity to slip into a more extreme mode of climatic variability, namely one in which megadroughts are more frequent and long lasting.[64] These modes in the past occurred in the absence of human-caused buildup of greenhouse gases, but seem to be associated with periods of warmer climate conditions. The work presented at the meeting speculated that given the association between past megadroughts and warmer climate, there exists a potential danger of slipping into a new megadrought era akin to the medieval megadroughts if the future climate is warmer due to the buildup of greenhouse gases.[65]

DROUGHT FORECASTS FOR THE UNITED STATES

Predicting the severity and duration of severe drought over a specific region is not yet possible more than a few months in advance because of the many factors that influence drought. Nevertheless, some modeling studies suggest that a transition to a more arid average climate in the American West, perhaps similar to conditions in precolonial North America, may be underway.[66] Some studies have suggested that human influences on climate, caused by emissions of greenhouse gases, may be responsible for a drying trend;[67] however, other studies appear to indicate an opposite trend or possibility (see above section on "Drought and Climate Change").[68] Whether future greenhouse gas-driven warming can be linked to La Niña-like conditions, or other phenomena related to the El Niño-Southern Oscillation, is unclear.

A likely consequence of higher temperatures in the West would be higher evapotranspiration, reduced precipitation, and decreased spring runoff.[69] These impacts would result from an "acceleration" of the hydrologic cycle, due to increased warming of the atmosphere, which in turn increases the amount of water held in the atmosphere.[70] A possible consequence is more frequent, and perhaps more severe, droughts and floods. However, these changes are unlikely to occur evenly across the United States. Observations of water-related changes over the last century suggest that runoff and streamflow in the Colorado and Columbia River basins has been decreasing, along with the amount of ice in mountain glaciers in the West, and the amount of annual precipitation in the Southwest.[71] Yet the understanding of hydrologic extremes, such as drought, is confounded by other effects such as land cover changes, the operation of dams, irrigation works, extraction of groundwater, and other engineered changes. Forecasting drought conditions at the regional scale, for example for river basins or smaller, is difficult because current climate models are less robust and have higher uncertainty at smaller scales.[72]

Even though forecasting drought at the regional scale is difficult, understanding potential changes in long-term trends is important for water managers at all levels—federal, state, local, and tribal. Water project operations and state water allocations are typically based on past long-term hydrological trends; significant deviations from such trends may result in difficult challenges for water managers and water users alike.[73]

CONCLUSION

Drought is a natural hazard with potentially significant economic, social, and ecological consequences. History suggests that severe and extended droughts are inevitable and part of natural climate cycles. Drought has for centuries shaped the societies of North America and will continue to do so into the future. Current understanding is that the physical conditions causing drought in the United States are linked to sea surface temperatures in the tropical Pacific Ocean. For example, the 2011 severe drought in Texas is thought to be linked to La Niña conditions in the Pacific Ocean. Increasingly, studies are projecting the long-term role that droughts may play in regional climate patterns. Nonetheless, available technology and science remains limited to forecasting specific drought a few months in advance for a region. The prospect of extended droughts and more arid baseline conditions in parts of the United States represents a challenge to existing public policy responses for preparing and responding to drought.

End Notes

[1] Edward R. Cook, et al., "North American drought: reconstructions, causes, and consequences," Earth-Science Reviews, vol. 81 (2007): pp. 93-134. Hereafter referred to as Cook et al., 2007.

[2] Donald A. Wilhite, et al., Managing Drought: A Roadmap for Change in the United States (Boulder, CO: The Geological Society of America, 2007), p. 12; at http://www.geosociety.org/meetings/06drought/roadmap.pdf.

[3] These are the categories used by the National Drought Mitigation Center (NDMC). The NDMC helps prepare the U.S. Drought Monitor and maintains its website.

[4] NDMC data collected since 2000. U.S. Drought Monitor at the NDMC in Lincoln, NE. See http://droughtmonitor.unl.edu/DM_tables.htm?archive.

[5] In some years or months, however, no part of the country was under extreme or exceptional drought. For example, from January 2000 through early April 2000, extreme or exceptional drought did not affect any portion of the country.

[6] NDMC, http://www.drought.unl.edu/DroughtBasics/WhatisDrought.aspx.

[7] Evapotranspiration may be defined as the loss of water from a land area through transpiration from plants and evaporation from the soil and surface water bodies such as lakes, ponds, and manmade reservoirs.

[8] Permanently arid conditions reflect the climate of the region, which is the composite of the day-to-day weather over a longer period of time. Climatologists traditionally interpret climate as the 30-year average. See NDMC, http://www.drought.unl.edu/DroughtBasics/ Whatis Climatology.aspx.

[9] For more on fire, see CRS Report RL30755, Forest Fire/Wildfire Protection, by Kelsi Bracmort.

[10] For example, the Palmer Drought Index has been widely used by the U.S. Department of Agriculture to determine when to grant emergency drought assistance. See NDMC, http://drought.unl.edu/Planning/Monitoring/ ComparisonofIndicesIntro/PDSI.aspx.

[11] The five key indicators include the Palmer Drought Index, the Climate Prediction Center soil moisture model, U.S. Geological Survey weekly streamflow data, the Standardized Precipitation Index, and short- and long-term drought indicator blends. For a discussion of drought indices, see NDMC, http://droughtmonitor.unl.edu/current.html.

[12] U.S. Drought Monitor, http://www.drought.unl.edu/dm/classify.htm.

[13] National Weather Service Forecast Office, Reno, NV, http://www.nws.noaa.gov/ climate/index.php?wfo=rev.

[14] National Weather Service Forecast Office, San Francisco Bay Area/Monterey, http://www. nws.noaa.gov/climate/ index.php?wfo=mtr.

[15] Also, while drought may affect one region at a given time, other regions may experience too much water, and possibly flooding, at the same time. Water-related disasters, drought and flood, may occur simultaneously in different regions of the country.

[16] See NDMC, at http://drought.unl.edu/DroughtBasics/PredictingDrought.aspx.

[17] Ibid.

[18] Ibid.

19 Cook et al., 2007.

[20] Hoerling, Martin and Arun Kumar, "The perfect ocean for drought," Science, vol. 299 (January 31, 2003), pp. 691- 694. Hereafter referred to as Hoerling and Kumar, 2003.

[21] Herweiger, Celine, Richard Seager, and Edward Cook, "North American droughts of the mid to late nineteenth century: a history, simulation and implication for Mediaeval drought," The Holocene, vol. 15, no. 2 (January 31, 2006), pp. 159-171. Hereafter referred to as Herweiger et al., 2006.

[22] Cook et al., 2007.

[23] Richard Seager et al., "Model projections of an imminent transition to a more arid climate in southwestern North America," Science, vol. 316 (May 25, 2007): pp. 1181-1184.

[24] Hoerling and Kumar, 2003.

[25] Cook et al., 2007.

[26] Proxies are indirect measurements typically used where direct measurements are unavailable. Tree rings can be used as a proxy for measuring dryness and drought. Similarly, ice cores from glaciers and polar caps can be used as proxies for measuring atmospheric temperatures and carbon dioxide concentrations from thousands of years ago.

[27] Cook et al., 2007.

[28] Herweiger et al., 2006.

[29] Ibid.

[30] The Carey Act, signed into law on August 18, 1894 (Chapter 301, Section 4, 28 Stat. 422), initially made available up to 1 million acres of federal land in each state, provided that the state met several requirements for the eventual development of water resources for reclamation. Some observers have suggested that the failure of the Carey Act to foster irrigation projects in all the land made available, compounded in part by the 1890-1896 drought, led to the Reclamation Act of 1902 and the emergence of the Bureau of Reclamation in the 20th century. (See Marc Reisner, Cadillac Desert (New York, New York, Penguin Books, 1986)).

[31] Under the Reclamation Act of 1902 the federal government constructed hundreds of dams, reservoirs, and related facilities to provide water to local farmers to "reclaim" the arid West.

[32] Fye, F., D. W. Stahle, and E. R. Cook, "Paleoclimate analogs to twentieth century moisture regimes across the United States," Bulletin of the American Meteorological Society, 2003, vol. 84, pp. 901-909.

[33] For example, one report showed that 42% of the area studied in the American West was affected by drought during the years 900 to 1300, versus 30% between 1900 and 2003, a 29% reduction in the average area affected by drought between the two periods. See Cook et al., 2007.

[34] U.S. Drought Monitor, February 15, 2011, http://droughtmonitor.unl.edu/Maps AndData/ MapArchive.aspx.

[35] U.S. Drought Monitor, February 16, 2010.

[36] See the U.S. Drought Monitor, Texas, on October 4, 2011http://droughtmonitor. unl.edu/archive.html.

[37] "Climate Abyss: Weather and Climate Issues with John Nielsen-Gammon," Texas Drought Update, March 23, 2012, http://blog.chron.com/climateabyss/2012/03/texas-drought-update/.

[38] Office of the Texas State Climatologist, "Texas Drought Officially the Worst Ever," August 4, 2011, http://tamunews.tamu.edu/2011/08/04/texas-drought-officially-the-worst-ever/.

[39] John W. Nielsen-Gammon, The 2011 Texas Drought: A Briefing Packet for the Texas Legislature, October 31, 2011, p. 29, http://climatexas.tamu.edu/files/2011_drought.pdf. Possibly the most severe Texas drought overall occurred from 1950 to 1957, and had substantial impacts on water supplies across the state because it lasted over many years. Because of the longevity and severity of the 1950s drought, municipal water supplies in Texas today are designed to withstand a drought of similar magnitude, according to the state climatologist. Long-term precipitation patterns in Texas are influenced by a configuration of sea surface temperatures known as the Pacific Decadal Oscillation (PDO). Similar conditions also prevailed from the 1940s through the 1960s, encompassing the Texas drought of record (1950-1957).

[40] For the contiguous United States, nearly 80% of the land area was affected by abnormally dry or drought conditions. U.S. Drought Monitor, August 14, 2012, http://droughtmonitor. unl.edu/MapsAndData/MapArchive.aspx.

[41] U.S. Drought Monitor, Drought Monitor Archive Tables, http://droughtmonitor.unl.edu/ dmtabs_archive.htm.

[42] Martin Hoerling et al., An Interpretation of the Origins of the 2012 Central Great Plains Drought, National Oceanic and Atmospheric Administration, Assessment Report: NOAA Drought Task Force Narrative Team, March 20, 2013, ftp://ftp.oar.noaa.gov/CPO/pdf/ mapp/reports/2012-Drought-Interpretation-final.web-041113.pdf.

[43] Ibid., p.1.

[44] Ibid., p. 4.

[45] Ibid., p. 10.

[46] Ibid., p. 22.

[47] For a more in-depth discussion of California water supply and issues affected by drought, see CRS Report IF00008, California Drought: Water Supply and Conveyance Issues (In Focus), by Betsy A. Cody.

[48] California is the country's largest agricultural producer in terms of cash farm receipts—accounting for 11% (nearly $45 billion) of the U.S. total in 2012 (the last year for which data are available [www.cdfa.ca.gov/statistics/]). California is also the largest producer of many crops.

[49] For more information about the hydrology and policy issues involved in the 2007-2009 California drought, see CRS Report R40979, California Drought: Hydrological and Regulatory Water Supply Issues, by Betsy A. Cody, Peter Folger, and Cynthia Brougher.

[50] California Department of Water Resources, California's Drought of 2007-2009—An Overview, September 2010, http://www.water.ca.gov/waterconditions/ drought/docs/ DroughtReport2010.pdf.

[51] http://cdec.water.ca.gov/cgi-progs/iodir/WSIHIST.

[52] Office of Governor Edmund G. Brown, Jr., "A Proclamation by the Governor of the State of California—Drought," http://gov.ca.gov/news.php?id=16997.

[53] For information on current water supply conditions and historical water allocations to federal water contractors in California, see http://www.usbr.gov/mp/PA/water/.

[54] The Delta smelt is a species of fish listed as threatened under the federal Endangered Species Act and as endangered under the California Endangered Species Act. Natural Resources Defense Council v. Kempthorne, No. 1:05-cv-1207 OWW GSA (E.D. Cal., December 14, 2007).

[55] A discussion of ENSO is provided in the text box on page 7 of this report.

[56] S. D. Solomon et al., "Contribution of Working Group I to the Fourth Assessment Report of the Intergovernmental Panel on Climate Change," 2007, Cambridge University Press, Cambridge, United Kingdom and New York, NY.

[57] Ibid., Summary for Policy Makers.

[58] C. B. Field et al., IPCC, "Managing the Risks of Extreme Events and Disasters to Advance Climate Change Adaptation," 2012, Cambridge University Press, Cambridge, United Kingdom and New York, NY, p. 170.

[59] C.B. Field et al., IPCC, 2012, p. 170.

[60] According to the report, confidence in the validity of a finding is based on the type, amount, quality, and consistency of evidence and on the degree of agreement. Confidence is expressed qualitatively: low, medium, high.

[61] C.B. Field et al., IPCC, 2012, p. 172.

[62] The IPCC report was referring primarily to hydrological drought, and not to the other factors such as water demand, water infrastructure, and other factors that enhance the effects of hydrological drought.

[63] Quirin Schiermeier, "Climate Models Fail to 'Predict' US Droughts," Nature, vol. 496, no. 7445 (April 16, 2013), http://www.nature.com/news/climate-models-fail-to-predict-us-droughts-1.12810.

[64] Edward R. Cook, "Megadroughts: The scary past told by tree rings and its implications for the future," Abstract 1797911 presented at 2013 Fall Meeting, San Francisco, CA, American Geophysical Union, December 11, 2013.

[65] Some scientists suggest that it is too early to tell whether drought conditions in California and the southwest indicate the presence of a megadrought, and that even the Dust Bowl period of the 1930s didn't qualify as a megadrought. However, California has experienced more severe and longer lasting droughts in the past. See Elizabeth Harball, "Researchers warn Californians of epic droughts in the state's distant past," ClimateWire, February 20, 2014, http://www.eenews.net/climatewire/2014/02/20/stories/1059994822.

[66] Richard Seager et al., "Model projections of an imminent transition to a more arid climate in southwestern North America," Science, vol. 316 (May 25, 2007), pp. 1181-1184.

[67] Tim P. Barnett, et al., "Human-induced changes in the hydrology of the western United States," Science, vol. 319 (February 22, 2008), pp. 1080-1082.

[68] C.B. Field, et al., IPCC, 2012, p. 170.

[69] Research results are emerging, however, that suggest that local and regional patterns of precipitation may be variable, and parts of a region or a state could receive higher precipitation than the current average, even if the overall trend over the broader area is towards less precipitation. See K. T. Redmond, "Climate Change in the Western United States: Projections and Observations," Eos Trans. AGU, 90(52), Fall Meet. Suppl., Abstract U11D-02, 2009.

[70] National Research Council, Committee on Hydrologic Science, Global Change and Extreme Hydrology: Testing Conventional Wisdom, Washington, D.C., 2011, p. 3.

[71] Ibid., p. 7.

[72] Ibid., p. 9.

[73] P.C.D. Milly et al., "Stationarity Is Dead: Whither Water Management?," Science, vol. 319 (February 4, 2008), p. 574.

In: Water Resources in the United States
Editor: Dianna Harrington
ISBN: 978-1-63482-681-5

Chapter 4

EMERGENCY WATER ASSISTANCE DURING DROUGHT: FEDERAL NON-AGRICULTURAL PROGRAMS*

Nicole T. Carter, Tadlock Cowan and Joanna Barrett

SUMMARY

Drought conditions often fuel congressional interest in federal assistance. While drought planning and preparedness are largely individual, business, local, and state responsibilities, some federal assistance is available to mitigate drought impacts. While much of the federal assistance is targeted at mitigating impacts on the agricultural economy, other federal programs are authorized to provide non-agricultural water assistance. Interest in these non-agricultural programs often increases as communities, households, and businesses experience shrinking and less reliable water supplies. Authorized federal assistance is spread across a variety of agencies, and each has limitations on what activities and entities are eligible and the funding that is available.

Rural Utilities Service (RUS): The U.S. Department of Agriculture's RUS provides grants and loans for rural water systems in communities with less than 10,000 inhabitants; its programs are for domestic water service, not water for agricultural purposes. Some of the programs are tailored to emergency situations, while others may prioritize loans and

* This is an edited, reformatted and augmented version of a Congressional Research Service publication, No. R43408, dated February 26, 2014.

grants for communities and households facing drought-related declines in water quantity or quality. As of mid-February 2014, around $1.3 billion in loans and $370 million in grants are available for rural community water and waste systems. While these funds are provided for assisting with rural water systems broadly, systems affected by drought may receive priority. Also for FY2014, the RUS Household Water Well System Grants had received $1 million in appropriations, and the Administration had reprogrammed to the RUS Emergency Community and Water Assistance Grants program $3 million for California's rural communities.

Bureau of Reclamation: Reclamation's authorities to assist with emergency water supplies and conservation stem primarily from the Reclamation States Emergency Drought Relief (RSEDR) Program. RSEDR consists of various authorities, including direct Reclamation water assistance to reduce drought losses, water contract authority, technical assistance, drought planning grants, and actions to facilitate water purchases and transfers. Reclamation can provide much of this assistance to water users (including municipalities and water districts), private entities, tribes, and states. Most of the RSEDR program's authority is limited to the 17 western states and Hawaii. RSEDR emergency actions often are provided by Reclamation at 100% federal expense, although some nonfederal reimbursement is authorized. These emergency actions are available to communities and water providers, regardless of their size, but are prioritized by need and congressional direction. As of mid-February 2014, RSEDR funding was $0.5 million for FY2014.

Army Corps of Engineers (USACE): The Corps has authority to assist emergency water supplies and their transport when state resources are exceeded and a public health threat is imminent. This authority has largely been used for assisting tribes with imminent drinking water supply issues. These activities have generally been funded through reprogramming of available agency funds. The agency also has authority to contract for provision of limited quantities of water (if available) from its reservoirs for municipal and industrial purposes.

Role of States and Other Federal Authorities. If a drought's effects overwhelm state or local resources, the President, at the request of a governor or tribal governing body, is authorized under the Stafford Act (42 U.S.C. 5121 et seq.) to issue major disaster or emergency declarations resulting in federal aid to affected parties. Since the 1980s, however, requests by U.S. states for Stafford Act drought-related declarations and related assistance for drinking water supplies have been denied. The U.S. Secretary of Agriculture has overseen most federal drought response through agricultural disaster assistance.

DROUGHT AND EMERGENCY WATER SUPPLIES: A PRIMER

Droughts raise several issues for Congress, including how to measure and predict drought, how to prepare, and how to coordinate federal agency actions responding to drought. In addition to resulting in agricultural losses, droughts also can affect the water supplies of individuals, communities, industries, species, and other services. Although many water allocation and other water management responsibilities lie largely at the state or local level, stakeholders may seek federal assistance with non-agricultural water supplies during droughts. Congress has created various programs to assist with non-agricultural water supplies for communities, households, and fish and wildlife habitat. These programs are spread across the jurisdictions of multiple agencies and congressional committees and vary widely in their use and funding. Current drought conditions, especially in western states, and exceptional and extreme drought conditions in portions of several states, including much of California, may test the effectiveness of these programs and raise questions about their role in federal drought response and policy.

FEDERAL DROUGHT AND EMERGENCY WATER ASSISTANCE

States, along with local governments and water providers, generally are responsible for preparing and planning for drought conditions. Often states take actions guided by state-level drought plans.[1] Some states also have state assistance and authorities that are used to alleviate the impacts of drought conditions. For example, some states (e.g., California, Idaho, and Texas) have instituted water banks and water transfer mechanisms to deal with water supply shortages.

The de facto federal policy since the 1980s has been that the U.S. Secretary of Agriculture takes the lead in responding and declaring eligibility for federal agricultural disaster assistance, including drought-related disasters. A declaration of an agricultural disaster area by the Secretary of Agriculture triggers the availability of multiple agricultural assistance programs, most notably the programs of the Farm Services Agency (FSA), and may trigger availability of other federal programs, such as the Economic Injury Disaster Loans of the Small Business Administration (SBA).[2] Most of these programs are focused on mitigating the economic impacts of drought on agriculture or water-dependent small businesses. The federal programs authorized to assist with non-agricultural water supply emergencies are implemented largely independently from these agriculture disaster programs at FSA and SBA.

The federal agencies authorized to assist with non-agricultural water supplies emergencies are:

- U.S. Department of Agriculture (USDA) Rural Utilities Service (RUS), through various water system loan and grant programs (FY2014 funds: $1.3 billion in loans, $374 million in grants). While most of these funds are provided for assisting with rural water and waste systems broadly, systems and households affected by drought may receive a priority.
- Department of the Interior's Bureau of Reclamation (Reclamation), through activation of the agency's Reclamation States Emergency Drought Relief Program assistance and contract authorities (FY20014 funds: $0.5 million).
- Department of Defense's U.S. Army Corps of Engineers, through its emergency drinking water and its water contract authorities (no funds specified for FY2014).

Additional funds may become available through reprogramming or supplemental appropriations. Some authorized drought assistance programs have never been implemented, while others are used in limited circumstances, and others operate almost annually. For example, while Reclamation's direct assistance with well construction is regularly used (albeit limited by available appropriations), the Corps' drought emergency water assistance has been employed in more limited circumstances, primarily for tribes, due to the high thresholds for its use and nonfederal funding requirements.

Figure 1 provides a summary of the RUS, Reclamation, and Corps programs and authorities and what triggers the availability of assistance. Numerous other federal programs and activities may assist in reducing long-term water use in urban, rural, agricultural, and industrial uses. These actions to promote water conservation and efficiency are beyond the scope of this report.[3]

Agency	Trigger for availability of program or authority	Geographic eligibility	Program	Eligible entities
USDA - Rural Utilities Service	Rural water assistance available through competitive application process (emergency situations may receive priority)	U.S.	Community Water and Waste System Grants and Loans	R
			Emergency Community Water Assistance Grants *(to secure water after decline of quality or quantity)*	R
			Household Water Well System Grants *(grants to nonprofits, lending to rural households)*	I
Bureau of Reclamation	Secretary of the Interior approves request by Governor for availability of the Reclamation States Emergency Drought Relief (RSEDR) program or automatically available within area with approved RSEDR drought contingency plan	17 western states and Hawaii	Reclamation Activities to Mitigate Drought Losses and Damage *(e.g., temporary works, wells, water purchases)*	T W
			Loans for Water Activities to Mitigate Drought Losses and Damages	T W
			Provisions or Purchase of Water to Mitigate Drought-Related Fish and Wildlife Losses	FW
			Temporary Water Contracts & Nonfinancial Assistance to Facilitate Water Purchases and Transfers	T W
U.S. Army Corps of Engineers (USACE)	Upon request by Governor, USACE identifies area as drought distressed	U.S.	Emergency Water Transport and Well Construction	T W F
			Temporary Contracts for Municipal and Industrial Water Withdrawal from Core Reservoirs	S W I P

TYPICAL FEDERAL FUNDING MECHANISMS

- Funding regularly received through annual appropriations, at times funds added through reprogramming and supplementals
- Funding principally through reprogramming and supplementals
- Never Funded
- Temporary and Emergency Activities (no fiscal assistance)

ELIGIBLE ENTITIES

- T Tribes
- S States
- W Communities, public or private water providers
- R Rural communities (less than 10,000 residents)
- I Individuals
- P Private concerns
- F Farm business, ranchers
- FW Fish and wildlife entities

Source: CRS. Figure is best viewed in color.

Figure 1. Select Federal Non-Agricultural Emergency Water Assistance Programs.

USDA Rural Utilities Service[4]

RUS provides grants and loans for rural community and household water. Some of the programs are tailored for emergency situations, while others may prioritize loans and grants for rural communities and households facing drought-related declines in water quantity or quality. For RUS programs, rural communities are often defined as those with populations less than 10,000. The RUS programs that may assist in addressing drought-related rural water issues include:

- Water and Waste Disposal Grants and Loans: This RUS program provides grants and loans for rural community water systems. RUS may choose to prioritize during its competitive selection process projects for communities affected by droughts. While this is a broad program not focused on drought or emergencies, some of its funds in FY2014 may assist drought-affected communities (FY2014 funds: $1.3 billion in loans and $370 million in grants available). While most of these funds are provided for assisting with rural community water and waste systems broadly, systems affected by drought may receive a priority.
- Emergency Community Water Assistance Grants: This RUS program provides grants specifically to rural water systems experiencing an emergency resulting from a significant decline in quantity or quality of drinking water (FY2014 funds: $3 million reprogrammed by Administration for use in California communities, no other funds available).
- Household Water Well System Grants: This RUS program provides grants to nonprofit organizations for operating lending programs for the refurbishing of household water well systems in rural areas. Loans are to be made to individuals with low or moderate incomes. Some of this program's FY2014 funds may assist drought-affected households (FY2014 funds: $1 million).

More information on these RUS rural water programs is available in the **Appendix**.

Bureau of Reclamation[5]

Reclamation's authorities to assist with emergency water supplies and conservation stem primarily from the Reclamation States Emergency Drought Relief (RSEDR) Program (43 U.S.C. 2201, et seq.). The RSEDR program consists of a number of different authorities, including direct Reclamation water assistance and loans to reduce drought losses through temporary measures (except for certain wells),[6] water purchases, temporary water contracts, drought planning grants, actions to facilitate water purchases and transfers, and technical assistance. Reclamation can provide much of this assistance to water users (including municipalities and water districts), private entities, tribes, and states. Most of the RSEDR program's authority is limited to the 17 western states[7] and Hawaii, with the exception being that technical assistance is available nationally. RSEDR emergency actions often are provided by Reclamation at 100% federal expense, although some nonfederal reimbursement is authorized. These emergency actions are available to communities and water providers, regardless of their size, but are prioritized by need and congressional direction. As of mid-February 2014, RSEDR funding was $0.5 million for FY2014.

Not all of the authorized activities have been implemented. For example, the authority for Reclamation to provide loans for nonfederal water entities to undertake activities to reduce losses and damages from droughts has never been used. Reclamation's drought relief program received $0.5 million in appropriations for FY2014.[8] More information on Reclamation's direct assistance is available in the **Appendix**.

Reclamation has other authorities to address variability in hydrology that are not specific to drought or emergency response. For example, Reclamation has authorities, which are now coordinated under the Department of the Interior's WaterSMART efforts,[9] to promote water resources management preparedness, sustainability, and water and climate variability resilience through improved knowledge and information and mitigation actions such as water and energy efficiency, conservation, and planning. These do not appear in **Figure 1**, since they are not designed to function as emergency response programs and were not targeted specifically at drought or drought emergencies and disasters. WaterSMART grants typically are available for water conservation and water and energy efficiency projects, including system optimization, and advanced water treatment. To date, most grants have been used for water reuse projects (e.g., recycling of urban wastewater),[10] irrigation districts' water efficiency improvements, water banking, and watershed basin studies and activities. These grants typically cover only a portion of project costs (25% to 50% of project costs).[11]

Army Corps of Engineers

The Corps (USACE) has authority to assist in the provision and transport of emergency water supplies when state resources have been exceeded and there is an imminent public health threat. While USACE is authorized to assist political subdivisions, farmers, and ranchers with non-irrigation water, this authority has largely been used for assisting tribes with drinking water supplies. These activities have generally been funded through reprogramming of agency funds. The agency also has authority to participate in temporary contracts to provide limited quantities of water (if available) for municipal and industrial purposes (33 U.S.C. 708).[12] More information on the Corps' provision and transport of emergency water supplies is available in the **Appendix**.

Federal Assistance in the Late 1970s and Today

During the 1976-1977 drought that affected numerous western states, Congress provided additional appropriations to multiple federal agencies to assist in responding to the drought's impacts. This funding consisted of $175 million (in 1977 dollars) for the Economic Development Administration (EDA), $130 million for the Bureau of Reclamation, and $100 million for agricultural assistance. While many of the activities that were funded by Reclamation and the agricultural assistance are eligible for assistance by current federal programs, there is no directly comparable program today to the EDA activities.

The EDA Community Drought Relief program targeted its grants and loans at communities and cities with populations over 10,000 people; its basic goals were to augment community water supplies, aid in purchase and transport of water, and promote water conservation. Reclamation can assist eligible communities of any size and other eligible entities in the 17 Reclamation states and Hawaii with qualifying emergency measures for wells, temporary works, and conservation measures through its Reclamation States Emergency Drought Relief Program, subject to the availability of appropriations ($0.5 million in FY2014).

The U.S. Environmental Protection Agency has no emergency water assistance program for drinking water systems experiencing drought conditions.

Source: California Department of Water Resources, *The 1976-1977 California Drought: A Review*, May 1978.

DROUGHT DISASTERS: FEDERAL DECLARATIONS AND PROGRAMS

Additionally, if the effects of a drought overwhelm state or local resources, the President, at the request of the state governor or tribal governing body, is authorized under the Stafford Act (42 U.S.C. 5121 et seq.) to issue major disaster[13] or emergency declarations[14] resulting in federal aid to affected parties.

However, requests by U.S. states for Stafford Act drought-related declarations and related assistance since the 1980s have been denied.[15] The infrequency of presidential domestic drought declarations increases the uncertainty about the circumstances under which such a declaration is likely to be made.

Therefore, the emergency assistance typically available to mitigate and respond to the impacts of drought on non-agricultural water supplies is primarily through the USDA's Rural Utilities Service, the Bureau of Reclamation's Emergency Drought Relief Program, and in limited circumstances the U.S. Army Corps of Engineers emergency water authorities. Since FEMA has deferred to USDA on drought declarations over the past three decades, there are no sections of the Stafford Act or regulations or policy guidance documents that address, or even appear to lend themselves to, a prolonged drought event as opposed to a disaster incident. On February 14, 2014, the President announced multiple actions to address drought conditions in California, including accelerating and reprogramming funds to assist with water conservation and irrigation using existing authorized programs; no Stafford Act declaration was made.

FEDERAL INFRASTRUCTURE AND DROUGHT INFORMATION

Both the Bureau of Reclamation and the U.S. Army Corps of Engineers also alter the operations of their water resources facilities in response to droughts. Operations of federal water resource facilities, particularly reservoirs behind dams, may help meet water supply needs during droughts, yet can also be vulnerable to the effects of droughts. Federal dams, particularly in the West, were constructed in part to provide multi-year storage to help with variations in seasonal and annual precipitation. Sustained hydrological drought nonetheless affects operations of federally managed reservoirs, dams, locks, hydroelectric facilities, and other components of the nation's water infrastructure. The extent to which operational changes are authorized and can be implemented to reduce the water supply impacts of drought often is limited by the federal infrastructure's role in meeting multiple project purposes and satisfying legal requirements.

Recent legislation has expanded Reclamation's authority for its drought-related operations. The Consolidated Appropriations Act of FY2014 (P.L. 113-76) expanded the Secretary of the Interior's authority to participate in nonfederal groundwater banking in California, and waived certain reporting provisions for transfer of irrigation water among selected federal water contractors, while also directing Reclamation and the Fish and Wildlife Service to expedite "programmatic environmental compliance" to facilitate Central Valley Project (CA) water transfers. The same legislation also provided direction to the Corps regarding operation of its facilities during drought. The explanatory report accompanying P.L. 113-76 directed the Corps to work with communities to increase water storage in winter months, without significantly increasing flood risk, to assist those facing drought conditions.

The federal government is also a significant provider of drought information. Congress enacted the National Integrated Drought Information System (NIDIS) Act of 2006 (P.L. 109-430), which established NIDIS within the National Oceanic and Atmospheric Administration (NOAA) to improve drought monitoring and forecasting abilities. NIDIS coordinates the collaborative Drought Monitor and other drought forecasting information, which are used by a variety of stakeholders to plan their actions and investments. The U.S. Geological Survey also provides drought-related water information through its hydrologic monitoring network.

This information is used by communities and states to plan water withdrawals and diversions, assess needs for water-use restrictions, and anticipate or respond to drought-related environmental stresses (e.g., fish kills, saltwater intrusion into aquifers, and habitat degradation from high water temperatures).

FEDERAL DROUGHT COORDINATION AND POLICY

During the widespread drought conditions of 2012 and early 2013, the National Disaster Response Framework was used by the Secretary of Agriculture to coordinate the federal drought response. The National Disaster Recovery Framework (NDRF) is the federal framework followed to assist disaster-affected communities to recover from a disaster.[16] An outgrowth of that experience was the creation in November 2013 of the National Drought Resilience Partnership to align federal drought policies across the government and to facilitate access to drought assistance and information sharing across all levels of government.[17]

While the partnership was anticipated to function largely as part of the President's Climate Action Plan to help prepare for drought, the western water supply conditions in 2014 have resulted in the partnership playing a coordinating role in federal drought response.[18]

The economic and social consequences of a water supply emergency can be costly to individuals, communities, businesses, and governments. Early actions may help avoid a non-agricultural water supply emergency resulting from drought. These actions can range from rehabilitating wells to improve their reliability, to reducing leaks in conveyance and distribution systems, implementing water rationing, and other water conservation measures. Consequently, a recurrent policy issue is how to effectively and efficiently mitigate drought's water supply impacts across the local, state, and federal agencies. How well current programs help to prevent drought conditions from becoming drought emergencies and drought disasters is one policy question. Another is whether the fragmentation of existing authorizations results in duplication, waste, gaps, and perverse incentives to prepare for drought. The broader policy struggle is how to transition from drought assistance to drought risk reduction.

APPENDIX. FACT SHEETS FOR FEDERAL PROGRAMS PROVIDING EMERGENCY WATER ASSISTANCE

Rural Utilities Service: Water and Waste Disposal Grants and Loans[19]

Form of Assistance, Condition of Assistance, and Eligibility Requirements

The Rural Water and Waste Disposal Account, administered by RUS, supports construction and improvements to rural community water systems (i.e., drinking water, sanitary sewerage, solid waste disposal, and storm drainage facilities). The program is not exclusively a drought program; however, water systems affected by drought may receive priority during implementation. Eligible rural areas may include an area in any city or town that has a population of less than 10,000 residents and where the projects are needed to meet applicable health or sanitary standards. Applicants must be unable to obtain sufficient credit elsewhere at reasonable rates and grants are made only if needed to reduce user charges to a reasonable level given the median household income for the area. Direct loans carry interest rates of 5% or less.

Required Approvals and Application and Selection Provision

The program is generally available and receives regular appropriations. No specific trigger is required for eligible entities to apply for assistance. Applications for direct loans, guaranteed loans, and grants are made with state Rural Development offices.

Authority and Appropriations

This assistance can be provided under the authority of the Consolidated Farm and Rural Development Act of 1972 (7 U.S.C. 1926). Funding for the program is allocated to state rural development offices where state directors set priorities particular to the rural needs of their state. Loan authorization level for direct loans is $1.2 billion; loan authority for loan guarantees is $50 million and loan subsidy levels for loan guarantees are $355,000 in FY2014. FY2014 annual appropriations provided $345.5 million for water and waste disposal grants. While most of these funds are provided for assisting with rural community water and waste systems broadly, those systems affected by drought may receive a priority. Funding also may be provided through reprogramming and emergency supplemental funding.

Rural Utilities Service: Emergency Community Water Assistance Grants[20]

Form of Assistance, Condition of Assistance, and Eligibility Requirements

RUS administers emergency and imminent community water assistance grants.[21] Eligible communities must face an emergency or a significant decline in quantity or quality of drinking water to apply. Grants can be made in rural areas and for cities or towns with a population of 10,000 or less, where the median household income is less than or equal to the state's non-metropolitan median household income. Eligible entities include most public bodies (e.g., state and local governmental entities), nonprofit corporations, and federally recognized Tribes serving rural areas. Privately owned wells are not eligible. Grants may cover 100% of project costs. Grants for projects where a significant decline in quantity or quality of water occurred within the last two years may not exceed $500,000 and grants for emergency repairs and replacement of facilities on existing systems may not exceed $150,000.

Required Approvals and Application and Selection Provision

Grants are only given for communities facing actual or imminent drinking water shortages. Applicants must submit an application, which then enters a nationally competitive process that evaluates the severity of the actual or imminent decline. The Secretary of Agriculture will act on all applications within 60 days of their submission to the extent practicable.

Authority and Appropriations

Assistance can be provided under the authority of Section 306A of the Consolidated Farm and Rural Development Act of 1972 (7 U.S.C. 1926a). At least 50% of funds provided annually must benefit rural communities with populations less than 3,000. The program is authorized for appropriations of $35 million annually in FY2014 through FY2018.

For FY2014, the program was appropriated no funds. In February 2014, it was announced that the program would receive $3 million in reprogrammed funds in FY2013 to be made available for California's rural communities (less than 10,000 residents with a median household income less than $62,883).[22]

Additional funds often become available through reprogramming and may become available through supplemental appropriations.

Rural Utilities Service: Household Water Well System Grants[23]

Form of Assistance, Condition of Assistance, and Eligibility Requirements

RUS provides grants to nonprofit organizations for lending programs to refurbish household water well systems in rural areas; loans are to be made to individuals with low or moderate incomes. Eligible individuals include members of households in which members have a combined income at or below the median non-metropolitan household income for the state or territory in which the individual resides, according to the most recent United States census. Loans made with grant funds must not exceed $11,000 per water well system and must have an interest rate of 1%. Financed facilities will not be inconsistent with any development plans of the state, multi-jurisdictional area, county, or municipality in which the proposed project is located.

Required Approvals and Application and Selection Provision

Nonprofit organizations must demonstrate expertise in well-water systems. Grants are determined by the Secretary of Agriculture.

Authority and Appropriations

This assistance can be provided under the authority of Section 6012 of the 2002 farm bill (P.L. 107-171) and Section 306D of the Consolidated Farm and Rural Development Act of 1972 (7 U.S.C. 1926e).

For FY2014, the program has been appropriated $993,000.

Bureau of Reclamation: Reclamation RSEDR Activities to Mitigate Drought Losses and Damage[24]

Form of Assistance, Conditions of Assistance, and Eligibility Requirements

The Reclamation States Emergency Drought Relief (RSEDR) Act authorizes the Commissioner of the Bureau of Reclamation to undertake (or contract for) construction, management, and conservation activities that mitigate losses and damages resulting from drought conditions.[25] Costs incurred by Reclamation are non-reimbursable (i.e., 100% federal). Construction activities are limited to temporary facilities, except certain wells to reduce losses and damages from drought can be permanent. Activities may occur within or outside of authorized Reclamation project areas. Eligible entities are water users,[26] tribes, and local entities in the 17 Reclamation states[27] and Hawaii.

Required Approvals and Application and Selection Processes

An eligible entity can apply for assistance if (1) the Secretary of the Interior has approved a request for RSEDR assistance by a governor or tribal governing body or (2) the state or tribe has a drought contingency plan that has been approved by Reclamation and transmitted to Congress.[28] Eligible entities can request Reclamation's assistance through the area, regional, or Commissioner's Office. The Regional Office Drought Coordinators prioritize requests; Reclamation's Drought Coordinator approves or disapproves requests based on congressional direction, available funding, need, and compliance with environmental laws. The Regional Director can provide the assistance in compliance with state and federal law.

Authority and Appropriations

These activities are authorized under the Reclamation States Emergency Drought Relief Act of 1991, as amended (43 U.S.C. 2221). P.L. 113-76 extended the authority through September 30, 2017. The authorization of appropriations (43 U.S.C. 2241) for this assistance fell under the broader Reclamation States Emergency Drought Relief activities, which was $90 million for the period FY2006 through FY2012.

As of mid-February 2014, RSEDR activities had received $0.5 million in appropriations for FY2014.

Army Corps Assistance: Provision of Emergency Drinking Water

Form of Assistance, Condition of Assistance, and Eligibility Requirements

The U.S. Army Corps of Engineers can construct wells and transport water to provide emergency drinking water during drought conditions.[29] Corps assistance is provided only to meet minimum public health and welfare requirements in the immediate future that cannot be met by state or local actions or through reasonable conservation measures.[30] Transport expenses are non-reimbursable expenses (i.e., 100% federal); the purchase or acquisition of the water and the storage facility at the terminal point and permanent water facilities are reimbursable expenses. This authority cannot be used for the provision of water for livestock, irrigation, recreation, or commercial/industrial use. Eligible entities are limited to drought-distressed political subdivisions, farmers, and ranchers.

Required Approvals and Application and Selection Provision

A governor, his/her representative, or the governing body of a tribe must make a written request for assistance to the Army Corps of Engineers. The USACE Director of Civil Works or the Assistant Secretary of the Army (Civil Works) makes the determination that an area has an inadequate water supply causing, or is likely to cause, a substantial threat to the health and welfare of the inhabitants of the area. If an applicant submits a request directly to USACE, the submission will be referred to the State Emergency Management Agency or equivalent office.

Authority and Appropriations

This assistance can be provided under the authority of the Disaster Relief Act of 1974 (33 U.S.C. 701n). Funding is provided through the USACE Flood Control and Coastal Emergencies Account, which often receives some appropriations through annual Energy and Water Appropriations acts, with the majority of its funds through supplemental appropriations. The Corps has authority to reprogram its civil works funds to accomplish work under this authority. In many years, the Corps does not use this authority. The Corps most often uses this authority to assist tribes with emergency drinking water issues.

End Notes

[1] The National Drought Mitigation Center posts online copies of drought management, mitigation, or response plans for states and localities, nationwide; see http://drought.unl.edu/Planning/PlanningInfobyState.aspx.

[2] For more information on these loans, see CRS Report R41309, *The SBA Disaster Loan Program: Overview and Possible Issues for Congress*, by Bruce R. Lindsay. Nonfarm businesses, agricultural cooperatives, businesses engaged in aquaculture of less than 500 employees and most private, nonprofit organizations affected by drought may qualify for

loans of up to $2 million to help meet financial obligations and operating expenses which could have been met had the disaster not occurred.

3 For a discussion of USDA programs assisting with agricultural water conservation, development, and use, see CRS Report R40763, *Agricultural Conservation: A Guide to Programs*, by Megan Stubbs. For a discussion of federal programs that assist with municipal water systems, including those that may assist with water reuse and desalination facilities, see CRS Report RL30478, *Federally Supported Water Supply and Wastewater Treatment Programs*, coordinated by Claudia Copeland.

4 For CRS assistance with RUS, contact Tadlock Cowan, Analyst in Natural Resources and Rural Development, 7- 7600.

5 For CRS assistance with Reclamation, contact Betsy A. Cody, Specialist in Natural Resources, 7-7229.

6 Construction activities are limited to temporary facilities, except that wells drilled to minimize losses and damages from drought can be permanent.

7 Reclamation is generally authorized to construct projects only in the 17 western states (Arizona, California, Colorado, Idaho, Kansas, Montana, Nebraska, Nevada, New Mexico, North Dakota, Oklahoma, Oregon, South Dakota, Texas, Utah, Washington, and Wyoming), unless otherwise directed by Congress.

8 Activities are funded as part of Reclamation's Emergency Drought Relief Program, which is typically a line item in Reclamation's annual appropriations, which typically are provided through the Energy and Water Development Appropriations Act, or through supplemental appropriations (e.g., $40 million provided by American Recovery and Reinvestment Act, P.L. 111-5).

9 Much of the program has its roots in the SECURE Water Act (Subtitle F, Title IX of the Omnibus Public Land Management Act of 2009, P.L. 111-11, 42 U.S.C. 10364).

10 Support for specific Reclamation water reuse and recycling projects, including certain pilot and demonstration projects, is authorized under Title XVI of P.L. 102-575, as amended. In recent years, requests for Title XVI funding has been requested and appropriated under the umbrella of the WaterSMART program.

11 In February 2014, the Administration announced $7 million for a Reclamation funding opportunity (with an invitation to apply directed at California water or power entities) for agricultural water conservation. Bureau of Reclamation, WaterSMART Program Funding Opportunities Fiscal Year 2014, http://www.usbr.gov/WaterSMART/docs/2014WaterSMARTFundingOpportunities.pdf; for more information, see http://www.grants search-grants.html?keywords=R14AS00021.

12 The agency has chosen to limit the authority of Corps District Engineers to enter into contracts authorized by 33 U.S.C. 708 to 99 acre-feet. Although not the subject of this report, the Secretary of the Army also has the authority to contract for on an interim basis the delivery of water for irrigation (43 U.S.C. 390); only water that was designated for municipal and industrial purposes which is not currently under contract is eligible for these interim contracts.

13 FEMA has published a listing of the factors at 44 CFR 206.48 that are considered when evaluating a major disaster request by a Governor.

14 Emergency declarations are intended to provide short-term emergency assistance to save lives and protect property and to lessen the threat of a more catastrophic event. Given the immediacy of the work under an emergency declaration, an abbreviated palate of assistance programs are available for emergency declarations compared to assistance available under a major disaster declaration. For example, while an emergency declaration can provide debris

removal assistance, it cannot offer any forms of permanent, infrastructure repair work. Similarly, while emergency temporary housing is available through Section 408 of the Stafford Act, other programs assisting families and individuals are not. FEMA has not published a list of factors for emergency declarations similar to the list that is available for evaluating a governor's request for a major disaster declaration.

[15] The last presidential drought or water shortage disaster declaration for a U.S. state was for New Jersey in 1980.

[16] The NDRF, which was completed in September 2011, identifies capabilities necessary to assist communities affected by a disaster to effectively recover, including, but not limited to, rebuilding infrastructure; providing adequate interim and long-term housing; restoring health, social, and community services; promoting economic development; and restoring natural and cultural resources. For more information on the NDRF as part of the national preparedness system, see CRS Report R42073, *Presidential Policy Directive 8 and the National Preparedness System: Background and Issues for Congress*, by Jared T. Brown.

[17] For more information on the partnership, see http://www.drought

[18] For more on the Administration's actions in response to the California drought, see The White House, "Fact Sheet: President Obama Leading Administration-wide Drought Response," press release, February 14, 2014, http://www.whitehouse.gov/the-press-office/2014/02/14/fact-sheet-president-

[19] For CRS assistance with RUS, contact Tadlock Cowan, Analyst in Natural Resources and Rural Development, 7- 7600.

[20] For CRS assistance with RUS, contact Tadlock Cowan, Analyst in Natural Resources and Rural Development, 7- 7600.

[21] Additional information is available at the USDA website http://www.rurdev.usda.gov/UWP-ecwag.htm.

[22] USDA, "Emergency Community Water Assistance Grants," press release, February 14, 2014, http://www.usda.gov/ wps/portal/usda/usdahome?navid=DISASTER_ASSISTANCE.

[23] For CRS assistance with RUS, contact Tadlock Cowan, Analyst in Natural Resources and Rural Development, 7- 7600.

[24] For CRS assistance with Reclamation, contact Betsy A. Cody, Specialist in Natural Resources, 7-7229.

[25] For additional information, see http://www.usbr.gov/drought

[26] "Water users" was not defined in statute. The agency has indicated that eligible entities include municipalities and water districts. While water users that are Reclamation water contractors receive some preference, other water users are also eligible.

[27] See footnote 7 for a list of these states.

[28] Entities with approved contingency plans include the Hopi Tribe, Hualapai Tribe, and Navajo Nation; the states of Hawaii, New Mexico, and Utah; and San Juan County, Utah.

[29] For additional information, see http://www.usace.army.mil/Missions /Emergency Operations/Drought.aspx and Chapter 6 of USACE, Engineer Regulation, ER 500-1-1, Emergnecy Employment of Army and Other Resources, available at http://140.194.76.129 /publications/eng-regs/ER_500-1-1/toc.htm.

[30] The Corps provision of emergency drinking water assistance for drought-impacts areas has been limited. Most of the Corps assistance under this authority during the 2000s was to assist Indian tribes (e.g., tribes in North Dakota in 2002/2004, and Navajo Nation tribes in Arizona in 2002, and Kickapoo tribe in Kansas in 2000).

INDEX

#

20th century, viii, 8, 47, 55, 58, 62, 66

A

access, 79
accounting, 20, 34, 39, 67
adaptation, 7, 62
Africa, 52
agencies, vii, viii, 2, 3, 6, 9, 13, 14, 17, 26, 41, 69, 71, 72, 76, 79
agricultural economy, viii, 69
agricultural producers, 5
agriculture, 4, 26, 36, 37, 45, 51, 55, 72
Alaska, 48, 56, 57
alters, 53
American Recovery and Reinvestment Act, 85
anthropologists, 46, 54
appropriations, vii, 2, 3, 4, 5, 7, 8, 10, 11, 13, 16, 17, 40, 70, 72, 75, 76, 80, 81, 82, 83, 84, 85
Appropriations Act, 20, 39, 85
Appropriations Committee, 18
aquaculture, 84
aquatic ecosystems, vii, 1, 16
aquifers, 27, 79
Asia, 52, 53
assessment, 27, 62
atmosphere, 46, 58, 63
atmospheric pressure, 53
attribution, 62
authority(s), 10, 15, 24, 25, 38, 39, 40, 55, 70, 71, 72, 75, 76, 77, 78, 80, 81, 82, 83, 84, 85, 86
authorized programs, 77

B

banking, 75, 78
banks, 71
base, 27, 52
beneficiaries, 15
benefits, 8, 9
bilateral relationship, 38
blends, 65
blogs, 22
Bureau of Reclamation, vii, 2, 3, 4, 10, 17, 18, 30, 33, 41, 66, 70, 72, 75, 76, 77, 78, 82, 85
businesses, viii, 69, 79, 84

C

carbon, 65
carbon dioxide, 65
cash, 60, 67
CFR, 85
challenges, 3, 6, 16, 27, 64

cities, 50, 61, 76, 81
citizens, 60
Civil War, 55
classification, 50
climate, viii, 7, 11, 45, 46, 47, 49, 53, 55, 58, 61, 62, 63, 64, 65, 67, 68, 75
climate change, viii, 7, 11, 45, 46, 47, 59, 61
Colorado River, viii, 10, 11, 12, 19, 22, 24, 28, 30, 31, 32, 37, 41, 42
commercial, 4, 84
Committees on Appropriations, 20, 39
community(s), viii, 3, 14, 50, 69, 70, 71, 74, 75, 76, 78, 79, 80, 81, 86
community service, 86
competition, vii, 1
competitive process, 81
competitiveness, vii, 1
complexity, 62
compliance, 20, 26, 31, 34, 36, 40, 78, 83
configuration, 66
conformity, 41
Congress, v, vii, 1, 2, 3, 4, 5, 6, 7, 9, 10, 11, 12, 13, 14, 15, 16, 17, 18, 19, 20, 39, 45, 47, 71, 76, 78, 83, 84, 85, 86
consensus, 49, 53
conservation, 2, 4, 5, 9, 13, 17, 19, 30, 31, 32, 70, 72, 75, 76, 77, 79, 82, 84, 85
conservation programs, 5, 17
conserving, 37
Consolidated Appropriations Act, 4, 15, 78
construction, 2, 5, 7, 8, 10, 12, 15, 16, 26, 41, 72, 80, 82
consumption, 31
contamination, 7
contingency, 83, 86
controversial, viii, 10, 11, 19
controversies, 10, 11, 25
convention, 21
cooling, 6
cost, 14, 41
crises, 3, 38
crop(s), 4, 67
crop production, 4
cycles, 20, 25, 34, 42, 64

D

damages, 8, 14, 75, 83, 85
dams, vii, 1, 3, 8, 10, 14, 26, 63, 66, 78
danger, 46, 51, 63
deficiency(s), 25, 46, 49, 50
deficit, 34, 39, 42
degradation, 27, 31, 79
Delta, 7, 11, 12, 16, 31, 32, 60, 61, 67
Department of Agriculture, 72
Department of Defense, 4, 72
Department of the Interior, 4, 30, 41, 72, 75
depth, 67
diplomacy, 38
diplomatic efforts, 25, 38
direct measure, 54, 65
directors, 80
disaster, 5, 14, 17, 70, 71, 77, 79, 85, 86
disaster area, 71
disaster assistance, 5, 14, 17, 70, 71
discharges, 41
disclosure, 16
displacement, 47
distribution, 24, 26, 36, 79
drainage, 30, 80
drinking water, 70, 72, 74, 76, 80, 81, 83, 84, 86
drying, 53, 63

E

economic activity, 60
economic development, 86
ecosystem, vii, viii, 2, 3, 5, 7, 9, 12, 16
ecosystem restoration, vii, viii, 2, 3, 5, 9, 12, 16
Ecuador, 53
El Niño, 53, 63
electricity, 9, 41
emergency, 2, 10, 13, 14, 17, 65, 69, 70, 72, 74, 75, 76, 77, 79, 81, 83, 84, 85, 86
Emergency Assistance, 14, 17
emergency declarations, 70, 77, 85
emergency repairs, 81

emergency response, 14, 75
employees, 84
encouragement, 55
endangered, vii, 2, 3, 7, 11, 31, 49, 60, 61, 67
endangered species, vii, 2, 3, 7, 11, 31, 49, 60, 61
energy, 4, 6, 7, 9, 26, 49, 75
energy efficiency, 75
environment, 3, 5, 12
environmental policy, 2, 4
environmental protection, 30, 31
Environmental Protection Agency (EPA), 9, 13, 15, 76
environmental stress(s), 79
Europe, 53, 62
evaporation, 49, 65
evapotranspiration, 49, 63
Everglades, 5, 7, 16
evidence, 37, 62, 67
execution, 24
executive branch, 24
executive function, 40
expertise, 38, 82
exposure, 37
extinction, 61
extraction, 63

F

families, 86
Farm Bill, 17
farmers, 4, 30, 39, 66, 76, 84
farmland, 9
farms, 61
federal agency, 47, 71
federal aid, 70, 77
federal assistance, viii, 69, 71
Federal Emergency Management Agency, 9, 18
federal government, 7, 15, 66, 78
federal law, 83
FEMA, 14, 77, 85, 86
financial, 8, 85
firearms, 2, 5
fires, 49
fish, 7, 10, 13, 49, 61, 67, 71, 79
Fish and Wildlife Service, 78
fishing, 26
flexibility, 25, 37
flood management, vii, viii, 2, 3, 7
flooding, vii, 1, 3, 8, 14, 28, 65
floods, vii, 1, 3, 8, 13, 14, 17, 63
force, 30
forecasting, 45, 52, 62, 64, 78
forest fire, 55
formation, 52
funding, vii, viii, 2, 3, 5, 12, 14, 15, 16, 17, 27, 69, 70, 71, 72, 75, 76, 81, 83, 85
funds, 10, 12, 15, 70, 72, 74, 76, 77, 80, 81, 82, 84

G

Georgia, 10, 57
global warming, 53, 58
governance, 5, 16, 24
governments, 8, 14, 24, 31, 79
governor, 70, 77, 83, 84, 86
grant programs, 72
grants, 15, 42, 69, 70, 72, 74, 75, 76, 80, 81, 82, 85
greed, 42
green alga, 49
greenhouse, 46, 58, 63
greenhouse gas(s), 46, 58, 63
groundwater, 5, 27, 34, 51, 58, 63, 78
GSA, 67
guidance, 7, 77
Gulf Coast, 7, 16
Gulf of Mexico, 14, 32

H

habitat(s), 5, 7, 13, 16, 27, 71, 79
harbors, vii, 1, 9
Hawaii, 48, 56, 57, 70, 75, 76, 83, 86
health, 80, 84, 86
Holocene, 65

homes, 9, 27
House, 5, 18
household income, 80, 81, 82
housing, 86
humidity, 52
hunting, 26
Hurricane Katrina, vii, 1, 8
Hurricane Sandy, vii, 1, 3, 8, 10, 14, 17
hurricanes, 17
hydrological conditions, 61
hydropower generation, vii, 1
hydropower projects, vii, 1

I

improvements, vii, 1, 36, 42, 75, 80
income, 81, 82
indirect measure, 65
individuals, 71, 74, 79, 82, 86
Indonesia, 53
industry(s), 36, 71
information sharing, 79
infrastructure, vii, 1, 2, 3, 5, 8, 11, 12, 14, 15, 18, 27, 32, 36, 67, 78, 86
injury, 25
interest rates, 80
intervention, 34
investment(s), vii, 1, 2, 3, 7, 8, 14, 15, 31, 34, 36, 78
irrigation, 7, 8, 10, 21, 33, 41, 42, 49, 55, 61, 63, 66, 75, 76, 77, 78, 84, 85
issues, vii, viii, 1, 2, 3, 5, 6, 11, 12, 14, 16, 17, 20, 21, 30, 33, 37, 39, 43, 45, 47, 67, 70, 71, 74, 84

J

jurisdiction, 14, 40, 41

L

lakes, 49, 65
laws, 40, 83
lead, 50, 71
leaks, 79
legislation, vii, viii, 2, 3, 4, 5, 6, 11, 12, 13, 15, 16, 20, 78
legislative responses, 22
lending, 74, 82
levees, vii, 1, 3, 8, 9, 14
litigation, 11
livestock, 30, 84
loan guarantees, 15, 80
loans, 15, 69, 72, 74, 75, 76, 80, 82, 84
local government, 9, 14, 15, 71, 81
longevity, 66

M

magnitude, 14, 66
major disaster declaration, 85
majority, 13, 14, 18, 33, 84
management, vii, 1, 2, 3, 4, 7, 8, 10, 11, 12, 14, 20, 26, 27, 31, 36, 37, 38, 42, 71, 75, 82, 84
mass, 59
matter, 24, 42
measurement(s), 61, 62
media, 33
median, 80, 81, 82
meter, 53
Mexico, v, viii, 17, 19, 20, 21, 22, 23, 24, 25, 26, 27, 28, 30, 31, 32, 33, 34, 35, 36, 37, 38, 39, 40, 41, 42, 43, 85, 86
Midwest flooding, 10
mission, 10, 12, 17
Mississippi River, vii, 1
Missouri, vii, 1, 2, 6, 7, 10
models, 64, 67
modifications, 29
moisture, 46, 49, 51, 52, 62, 65, 66
Montana, 17, 85

N

National Aeronautics and Space Administration, 9
National Research Council, 68

natural disaster(s), 13, 17, 37, 46, 54
Natural Resources Conservation Service (NRCS), 9
Navajo Nation, 86
negotiation, 34
NOAA, 66, 78
nonprofit organizations, 74, 82, 84
North America, 29, 42, 46, 47, 52, 54, 62, 63, 64, 65, 68

O

Obama, 43
officials, 27, 31, 38, 39
oil, 6
Oklahoma, 17, 49, 85
operations, vii, 1, 3, 5, 7, 10, 11, 13, 41, 64, 78
opportunities, 13
optimization, 75
oversight, vii, 1, 3, 4, 6, 7, 12, 13, 14, 16

P

Pacific, 52, 53, 59, 64, 66
palate, 85
permit, 25
plants, 9, 49, 65
playing, 79
polar, 65
policy, vii, viii, 2, 3, 4, 13, 14, 17, 45, 47, 49, 67, 71, 77, 79
policy choice, 47
policy issues, vii, viii, 2, 3, 4, 45, 67
ponds, 65
population, 7, 10, 36, 60, 80, 81
population growth, 7, 36
portfolio, 3, 6
power generation, 9
power plants, 9
precipitation, 11, 28, 38, 41, 46, 49, 50, 51, 52, 53, 56, 58, 60, 63, 66, 68, 78
predictability, 39
preparedness, viii, 3, 6, 13, 69, 75, 86
preservation, 49, 61
president, 38, 86
President, 24, 38, 40, 43, 70, 77, 79, 86
President Obama, 38, 43, 86
private sector, 50
project, vii, viii, 2, 3, 5, 7, 11, 12, 14, 15, 16, 17, 33, 61, 64, 76, 78, 81, 82, 83
project sponsors, 12
protection, vii, 1, 2, 3, 5, 7, 17, 31
public health, 70, 76, 83
public interest, 7
public policy, 46, 64
Puerto Rico, 48
pumps, 11

R

rainfall, 49, 53, 56, 58
reality, 32
reconstruction, 54
recovery, 12, 14
recovery plan, 14
recreation, 4, 7, 9, 10, 49, 84
recreational, vii, 1
recycling, 12, 75, 85
Reform, 2, 4, 13, 17
regions of the world, 62
regulations, 26, 77
reliability, 12, 79
relief, 28, 50, 75
repair, 86
requirements, 7, 26, 34, 37, 66, 72, 78, 84
resilience, 75
resolution, 21, 30, 37, 38
resource management, 3, 12
resources, vii, 2, 3, 7, 8, 10, 12, 27, 58, 60, 70, 76, 77, 86
response, vii, viii, 2, 3, 8, 11, 12, 13, 14, 17, 61, 70, 71, 75, 78, 79, 84, 86
restoration, 2, 5, 7, 12, 16, 31, 37
restrictions, 2, 5, 49, 79
rights, 20, 24, 33
rings, 46, 54, 65, 67

Rio Grande, viii, 6, 19, 20, 21, 22, 23, 24, 25, 26, 28, 31, 32, 33, 34, 35, 36, 37, 38, 39, 40, 42
risk(s), 6, 8, 10, 14, 37, 60, 78, 79
risk management, 8, 37
river basins, 6, 21, 64
river flows, 3
roots, 85
rules, 16, 24
runoff, 11, 28, 41, 62, 63
rural areas, 74, 80, 81, 82
rural development, 80
Rural Utilities Service (RUS), 69, 72, 74, 77, 80, 81, 82, 85, 86

S

safety, 18
salinity, 19, 26, 30, 32, 41
salinity levels, 26, 30
salts, 30
saltwater, 79
savings, 36
science, 2, 6, 64
scientific understanding, 46, 61, 62
scope, 17, 26, 30, 47, 72
sea level, 53
Secretary of Agriculture, 70, 71, 79, 81, 82
Senate, 5, 15, 18, 27, 40
services, 7, 71
shortage, 42, 86
Sierra Club, 41
simulation(s), 62, 65
small businesses, 72
social consequences, 79
society, 46, 47, 54
solid waste, 80
South America, 53
South Dakota, 17, 85
southwest region, viii
species, 2, 7, 8, 10, 12, 37, 60, 61, 67, 71
spending, 16
Stafford Act, 13, 70, 77, 86
stakeholders, 6, 27, 31, 33, 34, 36, 37, 38, 71, 78
state(s), viii, 6, 7, 8, 9, 14, 15, 19, 27, 28, 30, 31, 32, 38, 39, 40, 41, 42, 50, 54, 55, 56, 58, 59, 60, 61, 64, 66, 68, 69, 70, 71, 75, 76, 77, 79, 80, 81, 82, 83, 84, 85, 86
statistics, 67
storage, 2, 5, 11, 12, 13, 15, 20, 33, 36, 37, 41, 58, 61, 78, 84
storms, 3
subsidy, 80
surplus, 31
sustainability, 75

T

target, 20, 21, 34, 35
Task Force, 66
technical assistance, 8, 70, 75
technology(s), 6, 64
temperature, 52, 53, 56, 62
temporary housing, 86
tension(s), 10, 20, 26, 31, 34, 35, 61
territory, 40, 41, 82
Title I, 85
trade, 8, 53
trade-off, 8
transparency, 39
transpiration, 49, 65
transport, 6, 70, 76, 83
transportation, 16
treaties, 20, 21, 40
treatment, 6, 75
triggers, 71, 72

U

U.S. Army Corps of Engineers, vii, 2, 3, 4, 9, 72, 77, 78, 83
U.S. Department of Agriculture (USDA), 9, 65, 69, 72, 74, 77, 85, 86
U.S. Department of the Interior, 17
U.S. Geological Survey (USGS), 9, 27, 65, 78
U.S. history, 55
United Kingdom, 67

United Nations, 31
United States, i, iii, v, viii, 6, 8, 13, 19, 20, 21, 24, 25, 26, 27, 28, 30, 31, 32, 33, 34, 35, 37, 38, 39, 40, 41, 42, 43, 45, 46, 47, 48, 51, 52, 53, 54, 55, 56, 57, 58, 59, 62, 63, 64, 66, 68, 82
urban, 3, 27, 36, 72, 75
urban areas, 36

V

vegetation, 46, 49
vulnerability, 62

W

Washington, 17, 68, 85
waste, 70, 72, 74, 79, 80
waste disposal, 80
wastewater, 16, 75
water quality, 5, 7, 21, 26, 27, 30, 49
water resources, vii, 2, 3, 4, 6, 7, 8, 9, 10, 11, 12, 15, 27, 66, 75, 78
water rights, 2, 5, 8, 10, 33, 35, 36
water supplies, viii, 3, 5, 7, 8, 10, 11, 12, 28, 36, 45, 49, 66, 69, 70, 71, 72, 75, 76, 77
water supply augmentation, vii, 1
watershed, 75
waterways, vii, 1, 4, 9
web, 66
welfare, 83, 84
wells, 75, 76, 79, 81, 83, 85
West Africa, 62
wetlands, 9, 17, 31, 41
wetting, 62
White House, 43, 86
wildfire, 51
wildlife, 7, 10, 13, 49, 61, 71